THE CALCULUS EXPERIENCE

A Tale of Intuition and Rigor
(Vol. 1 Pre-Calculus)

VAIBHAV VARUN

INDIA • SINGAPORE • MALAYSIA

Notion Press

Old No. 38, New No. 6
McNichols Road, Chetpet
Chennai - 600 031

First Published by Notion Press 2019
Copyright © Vaibhav Varun 2019
All Rights Reserved.

ISBN 978-1-64733-684-4

This book has been published with all efforts taken to make the material error-free after the consent of the author. However, the author and the publisher do not assume and hereby disclaim any liability to any party for any loss, damage, or disruption caused by errors or omissions, whether such errors or omissions result from negligence, accident, or any other cause.

While every effort has been made to avoid any mistake or omission, this publication is being sold on the condition and understanding that neither the author nor the publishers or printers would be liable in any manner to any person by reason of any mistake or omission in this publication or for any action taken or omitted to be taken or advice rendered or accepted on the basis of this work. For any defect in printing or binding the publishers will be liable only to replace the defective copy by another copy of this work then available.

Contents

Wishes from Cosmonaut Salizhan Sharipov — 9
Foreword — 11
From the Desk of the CEO — 12
To Those Who Have Made It Possible — 13
Greek Alphabets — 15

Chapter 1/∞: This Thing Called 'Calculus': Introduction — 17
    1/∞.1 Shadow, Stick and the Earth — 17
    1/∞.2 Chess, Math and Music — 20
    1/∞.3 Logic vs. Tools — 21
    1/∞.4 A Guide to This Book — 22
    Diagnostic Test — 25

Chapter 1: The Holy Laws of Mathematics and How to Break Them: Indeterminate Forms — 27
    1.1 Division by Zero — 27
    1.2 Indeterminate Forms — 30
    1.3 Closure — 32
    Exercises — 34

Chapter 2: Our Place in the Universe and Mathematics: Coordinate Systems — 35
    2.1 Descartes and His Fly: The Cartesian Plane — 35
        2.1.1 Putting Things a Bit More Formally — 37
        2.1.2 Knowing How Far We Are: The Distance Formula — 39

2.2	Why I Call Descartes the "Buddha" of Math: Analytical Geometry	40
	2.2.1 Geometry	41
	2.2.2 Algebra	42
	2.2.3 Descartes, Geometry, Algebra and the Enlightenment	43
2.3	Why Would Newton Stay Behind: The Polar Coordinate System	47
	2.3.1 Polar Coordinates	48
2.4	Degree vs. Radian Measure	50
	2.4.1 The Radian Measure	51
2.5	Let Me Tell You a Magical Story	52
2.6	Closure	54
	Exercises	55

Chapter 3: Let's (not) Blame Euler for This: $0, 1, e, i, \pi$ — 57

3.1	The constant e	57
	3.1.1 Oh Dear, Bernoulli(s)!	57
	3.1.2 Jacob Bernoulli and Infinite Money	59
3.2	The Euler's Number: e	61
3.3	The Famous π	63
3.4	Imaginary Number i	65
3.5	The Most Beautiful Equation in Math	66
3.6	Closure	67

Chapter 4: Decoding the Mind of Nature: An Introduction to Functions — 68

4.1	Leibnitz and the Algebra of Thought	69
4.2	Lobachevsky-Dirichlet Definition of a 'Function'	73
4.3	Functions as Mapping from Sets to Sets	79
4.4	Implicit and Explicit Functions	81
4.5	The Tree of Functions	81
4.6	Closure	83
	Exercises	85

Chapter 5:	To Visually See the Algebra: Polynomial Functions	87
	5.1 "Dark Ages" and Polynomial Functions	88
	5.2 The Constant Polynomial Function	90
	5.3 Linear Polynomial Function	92
	5.4 Family of Even-Power Polynomial Functions	95
	5.4.1 Da Vinci, Quadratic Functions and Mathematics of Beauty	97
	5.5 Family of Odd Power Functions	104
	5.6 Closure	108
	Exercises	109
Chapter 6:	May the Force Be with You: Rational and Irrational Functions	111
	6.1 Hooke, Newton, Universe and Rational Functions	111
	6.2 Rational Functions of Type $\frac{1}{x^n}$, $n = 2k + 1$	114
	6.3 Rational Functions of Type $\frac{1}{x^n}$, $n = 2k$	117
	6.4 First Encounter with Mathematical Aliens: the Asymptotes	118
	6.5 Climbing Higher and Higher: Vertical Asymptotes	120
	6.6 Getting Close to the Ground: Horizontal Asymptotes	121
	6.7 Sharp Turn Ahead: Slant Asymptotes	123
	6.8 Irrational Functions and the Murder of Hippasus	125
	6.9 To Ruin Everything You've Learned So Far	129
	6.10 Closure	130
	Exercises	132

Contents

Chapter 7: To Map Atoms and Cosmos: Trigonometric Functions   135
- 7.1 The Magic of Proportions   135
- 7.2 Aryabhatta's Trigonometric Ratios   137
- 7.3 The Unit Circle – Birthplace of Trig-Functions   140
- 7.4 The Sine Function   142
- 7.5 The Cosine Function   144
- 7.6 The Tangent Function   146
- 7.7 Bunch of Formulas   148
  - 7.7.1 Basic Trigonometric Formulas   149
  - 7.7.2. Sum and Difference Formulas   150
  - 7.7.3 Double Angle Formulas   150
  - 7.7.4 Product to Sum formula   151
- 7.8 Closure   153
- Exercises   155

Chapter 8: Life Is All About Growth: Exponential and Logarithmic Functions   157
- 8.1 Payasam, Chess and Exponential Functions   157
- 8.2 Exponential Growth Functions   159
- 8.3 Exponential Decay Functions   161
- 8.4 Natural Base Exponential Function   162
- 8.5 Watchmaker, Laird and Logarithms   163
- 8.6 A Few Tricks in Logs   166
- 8.7 The Two Bases of Logs   167
- 8.8 Closure   169
- Exercises   170

Chapter 9: From Language to Grammar: Miscl. Topics in Functions   172
- 9.1 Moving in Pieces: Piecewise Functions   172
  - 9.1.1 Absolute Value Function   174

9.2 To Form New Functions		175
9.2.1 Algebraic Operations		175
9.2.2 Composition of functions		176
9.2.3 Inverse Functions		177
9.3 Closure		183
Exercises		184
Chapter 10: The Mathematical form of Art: Graphing Functions		186
10.1 Closure		186
10.2 Half-Knowledge		195
Exercises		197
References		*199*

Mr. Salizhan Sharipov (Colonel) is a legendary cosmonaut who has spent over 200 days in space. He has conducted two career spacewalks totalling 10 hours and 34 minutes. He also is an experienced Air force Pilot who specializes in flying MiG-21 and has over 950 flying hours. He has performed various experiments pertaining to Life sciences on International Space Station.

Mr. Sharipov has been conferred with the highest honour of the Russian Federation – 'Hero of Russian Federation' and NASA Space Flight Medal (1998).

Wishes from Cosmonaut Salizhan Sharipov

Calculus is an integral part of any endeavour in science. It is the backbone of theoretical basis of almost all modern technology – right from your cell phone to space travel. In particular, the elements of precalculus are necessary to be mastered before beginning the actual study of Calculus. I have had the opportunity of learning Calculus from great teachers and books in my college days.

The Calculus Experience is truly a different and innovative approach to understanding the core concepts and ideas in a simple and yet day-to-day language. The chapters are well connected and are a deviation from the traditional approach to mathematics which a lot of students find boring, difficult and challenging.

I congratulate the writer and the entire team for this marvellous piece of work and wish them all the best for the main and the next part of the book. It is my sincerest hope that everyone who reads the book will develop an eye to see the beauty that is Calculus.

Salizhan Sharipov

(Cosmonaut)

Foreword

Dr. T. RAMASAMI
Formerly Secretary to Govt. of India
Department of Science & Technology
Ministry of Science and Technology
Nayudamma-Abdul Wahid Chair Professor

Department of Leather Technology
ANNA UNIVERSITY
Centre for Academic and Research Excellence
CSIR CENTRAL LEATHER RESEARCH INSTITUTE
Adyar, Chennai 600020
E mail : samisrisailam@gmail.com
Mobile : 098182 28329
Phone : 044 2443 7418

19.11.2019

Foreword

I am delighted that a book on calculus is being brought out with an objective of redressing the mathematics phobia observed among the current day youth. In the backdrop of some original contributions to the domain of calculus emanating from Baskara from Kerala, one could not help feeling the need for pedagogy innovations for helping mathematics education. Our current teaching methodologies seem to make natural talents shy away from the study of mathematics. Disconnect between methods of teaching and joy of learning mathematics.in our school and college systems may well be a cause for the observed reality. Abstraction without connections to real life systems does not seem to enable internalization of learning. High focus on examination with less stress on education is not an ideal equation for learning mathematics, the queen of all sciences

The idea behind the declaration of 22nd December, the birth anniversary of Srinivasa Ramanujan as the National Mathematics day is to galvanize the academic community to foster native abilities in mathematics in in an organized manner. To make learning of calculus joyful, we need novel pedagogy and inspirational learning materials and enthusiastic participation of educators. Pedagogy Research Initiatives in Mathematics Education (PRIME) is a need of the hour. I see this book as one such initiative.

The approach seems to have built on the similarities of learning of mathematics and languages. This is consistent with the emerging concepts in cognitive theories of learning mathematics languages and music. I see the book as an attempt to bring back the fun of learning mathematics with humour playing a happy medium.. Anecdotal format of communication for delivering concepts, supplementing abstraction with art forms of description and teaching essential concepts ahead of doing calculus are new aspects of this effort. I admire the considerable level of innovativeness deployed in pedagogy.

I congratulate the laudable effort and hope that it would prove a new approach for teaching mathematics.

From the Desk of the CEO

One of the central elements of Space Kidz India has been to create a sense of excitement about space science in kids that they do not get traditionally in schools and colleges. One can only imagine how easy it is to get lost in the pursuit of grades and senseless competition that is simply turning the current youth in mechanised robots.

The theoretical backbone of everything is space science and astronomy can be traced back to Calculus. Thus, this book is been written keeping our core values in mind. The idea that if there is no excitement, if you are not enjoying the subject and that if there is no sense of awe- you are doing it wrong.

The Calculus Experience – Vol. 1 is a collective output of everyone involved here at Space Kidz India. We have done crazy experiments – such as introducing memes to explain certain concepts, humor, use of day to day language and lot more – every single element of this book sets it apart from the traditional setup. I would like to congratulate the team for their innovative effort and hope that our future generations don't fear mathematics.

Dr. Srimathy Kesan
Founder, CEO
Space Kidz India

To Those Who Have Made It Possible

The Calculus Experience – A tale of intuition and rigor Vol. 1 (Precalculus) would have been impossible to write without the love, support and continued enthusiasm of a lot of people.

My father was the first person to introduce me to mathematics – and most importantly – to be original at it. I have always seen him in between books and naturally studying would not have to be forced upon me – I simply copied what he was doing anyway. My mother, on the other hand, made sure that she stayed up all along till I studied in the night. Amidst all the chaos and workload – the chilled-out part of me, that worries about nothing comes straight up from her.

To Mr. Salizhan Sharipov for conveying his best wishes for this book. Mr. Salizhan is a retired cosmonaut and has been to space two times. His two-career spacewalks have lasted more than 10 hours. Thank you for believing in this work!

To Padma Bhushan Dr. T. Ramasami for writing the foreword for this book. It is an absolute honour for all of us to receive your blessings and encouragement.

To Dr. H.C. Verma (retd. Prof. IIT, Kanpur), who has been a guiding force for quite some years now. I would like to thank you for all the help and support that you have provided.

To Varun Grover (Stand-up comedian & National-award winning lyricist) and his friends Pavan Jha, Subrat Mohanty and Svetlana Naudiyal. I cannot thank you guys enough for the interest, investment and encouragement that you have shown to my work without even personally knowing me.

To Dr. Srimathy Kesan (CEO and founder of Space Kidz India), the iron lady behind this project. Thank you for meeting all my long demands, following up on almost daily basis and being more excited about this project than I could be. It would have been impossible to do this without you.

To Kristin from Desmos, thank you for giving me permissions for using Desmos as a graphing calculator for this book. Desmos comes with an amazingly interactive set of tools that is a dream of for any mathematics student.

To various facebook pages like *Life through mathematicians' eye* and others who have let me use a lot of memes from their page. Thanks for the adding to the humour.

To Shivarama Krishnan for the sketches and Gobinath for the coordination – you guys have been amazing to work with. Thanks for meeting all my last-minute demands.

To my friend and brother Mr. J. Kokulkrishnan. Thank you would be small word for managing a hundred things for me and making all my work possible.

To a lot of people whom I have missed mentioning here – I thank you all again for keeping up with a maniac, hopelessly curious and an obsessively mad person like me. A lot of students, teachers and other people who have attended my seminars all these years and have helped me improve – thank you all.

Vaibhav Varun
Author

Greek Alphabets

It so happens that mathematics and sciences use a lot of alphabets from the Greek language – although, really, they are just alphabets – like of any other language and just contribute to the complicated look of the subjects. Therefore, they are presented right here in the starting with the message that you need not fear them.

Greek alphabet

Alphabet	Uppercase	Lowercase
Alpha	A	α
Beta	B	β
Gamma	Γ	γ
Delta	Δ	δ
Epsilon	E	ε
Zeta	Z	ζ
Eta	H	η
Theta	Θ	θ ϑ
Iota	I	ι
Kappa	K	κ
Lambda	Λ	λ
Mu	M	μ

Alphabet	Uppercase	Lowercase
Nu	N	ν
Xi	Ξ	ξ
Omicron	O	o
Pi	Π	π
Rho	P	ρ
Sigma	Σ	σ ς
Tau	T	τ
Upsilon	Y	υ
Phi	Φ	φ
Chi	X	χ
Psi	Ψ	ψ
Omega	Ω	ω

1/∞ This Thing Called 'Calculus': Introduction

"Why do children dread mathematics? Because of the wrong approach. Because it is looked at as a subject."

— *Shakuntala Devi*

Usually mathematics textbooks do not start like this, but here it goes – *I was never a mathematics person*. There, I said it! For me, mathematics was just another subject that I needed to pass the exams. That's all. Nothing more. It is how society and teachers project mathematics to you – collection of huge number of formulas, equations, proofs and everything else that we shall never use in our lifetime – which sadly is somewhat true. When you open a math book, written by people who have all sorts of academic degrees and teach you all sorts of boring unnecessary formulas (of course, with the "good" intention of teaching you math), it all makes no sense. The funda of our education system is pretty simple – you score good marks in exams; you are a good student and you shall progress in life. Wish it was that simple.

Having said that, all of us have this moment of awakening. That one time where we may not know what we should be doing, but we do realize what we should not be doing. One such moment happened to me while I was in Grade 9th, watching the very first episode of my all-time beloved television series – 'Cosmos' by Carl Sagan.

And I realized then the biggest blunder that I have been doing since my childhood – treating mathematics as a **"subject"**.

1/∞.1 Shadow, Stick and the Earth

The very first episode of Cosmos, titled "The Shores of the Cosmic Ocean" had this little segment about a man named Eratosthenes. He lived in the city of Alexandria, Egypt at around 240 B.C. He was the chief librarian of the Library of Alexandria. And this was a time when there was no way of communicating things around except for writing them down and passing them to different cities, thus exchanging information. Almost everyone believed that we are the center of the universe and that the earth is a huge flat surface. Everyone, except for one guy.

Eratosthenes was reading a book written by Pirate. Perhaps, he was passing time. He came across this page in the book that said that in the city of Syene on the day of Summer Solstice (21st June, the longest day of the year) at noon, a stick on the ground would cast no shadow. Hmm. Now anyone reading this would have dismissed it as yet another interesting but useless fact. However, Eratosthenes was a mathematician. He decided to repeat this observation in his own city.

And so, on the 21st of June, next summer, he placed a stick in the ground and waited to see no shadow of the stick as the exact time was nearing. However, something strange happened. The stick was actually making a shadow at the exact same time. 'How could it be?' he wondered that the same stick makes no shadow in Syene but casts a shadow in Alexandria on the exact same date and time. This fact troubled him. On a flat earth, indeed this is not possible. Since the sun is so far that all rays coming from it are parallel to us, all the sticks at the same day and at the same time must cast the same kind of shadows, he argued.

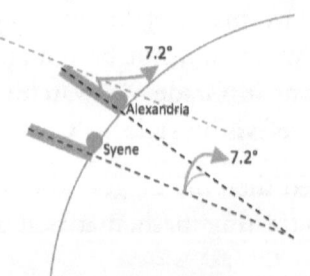

Fig. 1: Shadows of sticks

However, this uneasy feeling pointed towards only one thing – *that perhaps the earth is not flat, but rather spherical.* He measured that it the shadow of stick made an angle of 7.2° with the rays of the sun. The rules of geometry and parallel lines suggested that the sticks in

Alexandria and Syene extended backward all the way back to the centre of the earth should also make an angle of 7.2° as shown in Fig. 1.

He hired a man to measure the distance between Alexandria and Syene. It turned out that it was 5000 stadia or roughly 800 km. Now 7.2° is almost $1/50^{th}$ of 360°, so by ratio and proportions, the entire Earth's circumference should measure 50 times of 800 km i.e. 40,000 km. By the way, the actual circumference of the Earth is 40,075 km! Look at the remarkable accuracy here! An error of less than 0.2%.

What I essentially saw that someone 2000 years ago used nothing but sticks, shadows and the concept of parallel lines (which I learned in Grade 7th) to measure the entire circumference of the Earth – without even going around it. If that doesn't stun you, I don't know what probably will! How is it that when we studied about parallel lines, we never thought of this? That's when it occurred to me that mathematics isn't something that should be just taught in the class. It is perhaps *how nature speaks*.

I mean look around you – you'll find patterns everywhere. The shape of the trees, rivers, thunder, your own nervous system follow a branching-like pattern. It is perhaps as if nature wishes to talk to us in these symbols and patterns. And this language of nature is what we call mathematics.

And there was the correction of the blunder I mentioned – mathematics is not supposed to be taught as a subject, but rather a *"language"*.

The only difference being that mathematics isn't a language that humans have invented. We have, of course, given it our own set of symbols – but yet the "feel" of it comes straight from nature – it is a universal language after all.

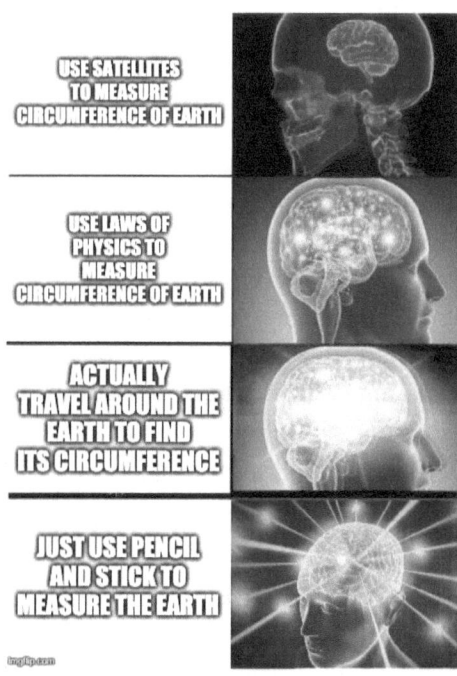

Meme 1: Mind=blown

And that is one of the main agendas of this book — to teach you how to speak "mathematics" and in particular, the *language of change* in nature or what we would like to call as *Calculus*.

1/∞.2 Chess, Math and Music

Anthony Braxton says *that playing chess, doing mathematics and playing music requires the exact same set of skills.* Sounds weird. After all, they are all so different. But let us have a deeper look at it.

Chess is played with a number of pieces — king, queen, bishops, pawn etc. These pieces do not mean that there is actually a king or queen there — but they are merely *representations* of them. Every piece moves according to a certain rule. Using just these pieces and rules, we can play a huge number of game variations — as long as we stick to the rule. Music also has its own language — the language of notes and scales. It has got its own rules too. And using those limited notes and scales — we have all the music in the world. Mathematics is same like chess and music. It has its own set of symbols and rules. Variables, numbers and operations like additions, exponents etc. Obviously learning rules is not fun in either chess or music — but they are necessary to get started; the same in mathematics.

Chess is a game of strategy. Before making any move, you must already have imagined the consequences of it. Every single move changes the entire game. So every move must be made while keeping the next move already in mind. Music too, is a business of disciplined strategy. While you play the current note, the next note must already be ready in your mind. You cannot just play random notes and expect it to give some kind of beautiful music. The same is in mathematics. It is a game of strategy — for every operation that you do, you must have already imagined the consequences of it. That's where the secret to doing good mathematics lies in. Understanding what each operation will do to the whole problem in hand.

Music and Chess both develop greatly with the mistakes of the past. The more you've played a particular piece of music — the more easily it comes to you. The more number of times you have played a game of chess, the easier it is for you to understand where a new game is going and who is going to win. The same stands for mathematics. Practice plays a pivotal role. The more you've done a certain amount of strange techniques — the more comfortable they are to use in any given problem.

And lastly, the last and my favorite argument – a sense of *'pleasure'*. A lot of people listen to or play music because it gives a sense of pleasure and beauty. Music appeals even to a person who has had no professional training in it – just because he/she experiences a sense of joy and pleasure. Experienced chess player, even after winning thousands of game are still intrigued by every new person they have to face-off with. And the same is true for mathematicians. Mathematics has a certain sense of 'pleasure' to it. We don't do it with a purpose in mind – like it is with music. It's done because it makes us happy. It is this pleasure of mathematics that I wish to convey with this book. I hope you'll achieve it.

So, if you feel you are not up for mathematics at this moment of time, take my crazy advice and go for learning either chess or any musical instrument!

$1/\infty.3$ Logic vs. Tools

This is a pre-calculus textbook – so of course it is going to present a lot of tools. It is essential that you master these categories of tools before actually pursuing Calculus in Vol. 2. But just because you know the tools doesn't mean you'll apply them without thinking in a particular problem. Here is an example I often give to Grade 6th:

Suppose there are 10 people available to work on an agricultural field and the work gets finished in 10 days. Now let us say that there are 20 people available to work on the same agricultural field. The question is: In how many days will the work be finished? The students immediately use their knowledge of ratio and proportions. They say now that we have double the people, work will be done in half the time and hence the work will get over in 5 days. So far so good. They have used the tool of ratio and proportion quite correctly.

Now we rephrase the question. A.R. Rahman composes a new song that is takes 6 minutes. He has 7 people working with him in the recording studio playing different instruments and taking care of the recording. After a few days, he is asked to perform the same song with an orchestra for an award function. The orchestra that he is playing with has 14 members in it who will handle different instruments. The question is: In how many minutes will the song be finished now? You'll be surprised that most of the students in class shout 3 minutes! Why? The same old application of ratio and proportions without seeing the scenario in hand.

The catch being that although they had the correct tool in mind – their logic was flawed.

That is one key point to note – knowing the tools and knowing where to use them are two different things. You need to think before applying any tool that you'll be learning in this book (well, basically anywhere). Try to understand the context of the problem you are solving rather than blindly applying the tools.

Meme 2: That one negative sign that ruins everything

1/∞.4 A Guide to This Book

If you are hoping that this book will help you crack any set of exams or something like that, then I am sorry to disappoint you. This text was never written keeping that agenda in mind. There is a saying that *'If you haven't come across a book that you want to read, then write it'*. That is the motto behind it. Growing up, I have felt that most textbooks are very mechanical in nature. They all tell you that they can teach you mathematics. They give you like a thousand problems to solve and what not. They do everything – except to teach you how to speak in mathematics!

The Calculus Experience – A tale of intuition and rigor (Vol. 1) is a small attempt at doing so. It is aimed at bringing a fine balance between intuition and rigor. Instead of randomly throwing in facts here and there – the goal is to give you a big picture. How a particular thought came into being, who were the people behind it and most importantly – how you can see mathematics as a completely intuitive thing and yet be able to understand the rigor of it. Being able to understand and spot patterns is one of the things that defines us as the most intelligent species. Definitely, in doing so, I have compromised on the standard hard-looking nature of mathematics – but so be it. Silvanus Thompson, in his famous book "Calculus made easy" has written:

"Considering how many fools can calculate, it is surprising that it should be thought either a difficult or a tedious task for any other fool

to learn how to master the same tricks. Some calculus-tricks are quite easy. Some are enormously difficult. The fools who write the textbooks of advanced mathematics—and they are mostly clever fools—seldom take the trouble to show you how easy the easy calculations are. On the contrary, they seem to desire to impress you with their tremendous cleverness by going about it in the most difficult way. Being myself a remarkably stupid fellow, I have had to unteach myself the difficulties, and now beg to present to my fellow fools the parts that are not hard. Master these thoroughly, and the rest will follow. What one fool can do, another can."

And that ladies and gentlemen – is the whole feeling behind writing this book. We are all fools, living for a small period on this planet – it should not be hard to master what other fools have mastered.

There are various **illustrations** spread across each chapter. The purpose is to teach you a concept that talking about it or to elaborate a concept that we have just learned. It is important to not miss reading them. The reason I have given them as illustrations and not examples is because of the informal space that it gives – a lot of space for talking to you without the fancy language of mathematics.

Exemplum serve as a demonstration to a particular formula or concept that we would have discussed. They are more formal in nature and are meant to show you how formal people in mathematics expect you to do certain questions.

There are various questions named as **Query** scattered here and there in each chapter. They usually ask you things which can serve as a further discussion on the same topic. It is important that you attend to each one of them.

Then, there are **Exercises**. Unlike most textbooks, I don't intend to throw you 1000 problems of the same kind to keep solving forever. In fact, most chapters will hardly contain any more than 20 problems in the exercises. However, I would advise you to think about each problem in hand and if you are not able to solve, you can re-read the chapter not only from this book but from any source possible. It is the concept that is of prime importance and not the source. Avoid looking up for the answers online or asking people. Doing so will only give you the answer to that very question – however, looking for 20 different sources and reading them till your concepts are clear and you are able to solve the problem is a whole different magic.

Finally, don't be in a rush to finish any topic or this book. Our current mathematical knowledge is the accumulation of hundreds of years of pursuit by thousands of people coming from all parts of the world. We certainly cannot do it all at once. Any subject is best conveyed by story and hence I have made sure to begin all chapters with some historical notes and incident.

$$(1.00)^{365} = 1.00$$
$$(1.01)^{365} = 37.7$$

Doing nothing at all
Vs.
Small consistent effort

Meme 3: Consistence

There is a Diagnostic Test at the end of this Chapter. This is to give you a fair idea of what skills you will be needing before you begin this book. However, do not be disappointed if you cannot solve them all. Look up, research and I am sure you'll be able to get them.

This precalculus text is more of a warm up for the Vol. 2. A fair warning is that the last few chapters will look technical and that's because they are. But sail through them. Vol. 2 is where you'll be applying all that you have learned. So take your time and be crystal clear in what you study here. I won't say it is going to be easy but I certainly hope that you discover the pleasure in the hard work – that whether you do end up scoring high marks in exams or not, if you just feel astounded by the sheer beauty of mathematics – at the end of the day, my purpose is served.

In the words of Spock, "Live long and Prosper". I hope you learn to understand the way nature speaks. Best wishes.

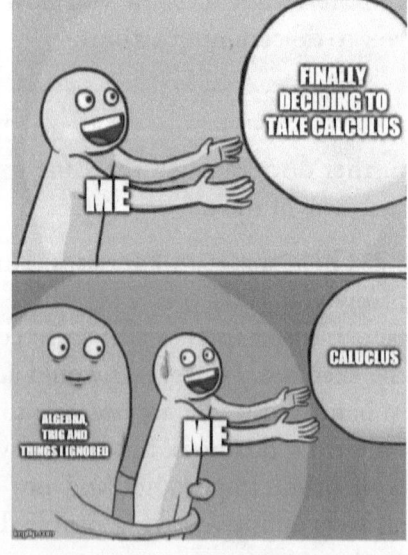

Meme 4: Why do you need Precalculus

Diagnostic Test

1. Draw the following intervals on a number line:
   a. $(3, \infty)$
   b. $(-4, 8]$
   c. $(-5, -1)$
2. Draw $\sqrt{2}$ on a number line.
3. Simplify the following fractions:
   a. $\dfrac{\dfrac{1}{x} - \dfrac{1}{y}}{\dfrac{1}{x} + \dfrac{1}{y}}$

   b. $\dfrac{5+\sqrt{2}}{\sqrt{7}-\sqrt{3}}$

   c. $\dfrac{xy}{1+\dfrac{1}{1+\dfrac{1}{xy}}}$

   d. $\dfrac{3}{x^2-16} - \dfrac{5}{x-4} + \dfrac{7}{x+4}$

   e. $\dfrac{1}{\dfrac{1}{x+a} + \dfrac{1}{x+b}}$

4. What does the Σ notation stand for? Express $4 + 7 + 10 +...$ in Σ notation.
5. Take the common factor out of the following expressions:
   a. $a^4 b^{10} + a^{60} b^{55} + a^3 b^{40}$
   b. $\gamma^{5/2} \beta^{3/4} + \gamma^{3/2} \beta^{7/4} + \gamma^{23/2} \beta^{31/4}$
   c. $a^2 b + b^2 c + c^2 a$
6. Which one of the following is greater?
   a. $\sqrt{2}$ or $7/5$
   b. $-\sqrt{10}$ or $-\pi$
   c. $1.3333...$ or $\dfrac{\pi}{\sqrt{64}}$

7. What is the difference between median and altitude of a triangle?
8. If $\dfrac{a}{b} = 7$, then what is the value of $\dfrac{\sqrt{3}b}{a}$?
9. If the ratio of is $a:b$, the ratio of $b:c$ is $3:2$ and ratio $c:d$ of, is $1:7$, then what is the ratio $a:d$?
10. How many times is $4\sqrt{3}x^5$ greater than $x^{3/2}$?
11. Is the sum, product and difference of two irrational numbers always irrational? Give examples.
12. Describe properties of the angle subtended by chord at any two points on the circle? What is meant by tangent and secant to a circle?
13. Can the numeric value of surface area and volume be same for any three dimensional shape? Give examples.
14. Suppose a pair of linear equations in two variables have no solutions to them. What changes would you make to one of them so that they now have infinite solutions?
15. Simplify the following exponents in the simplest form (Assume that all variables are positive):

   a. $\dfrac{5\sqrt{x} \cdot \sqrt{x^9}}{\sqrt[3]{750x^{11}}}$

   b. $\left(\dfrac{2a^5 b^2 c^{-11}}{7a^{3/2} b^{-11/3} c^{5/2}} \right)^{-2}$

   c. $\left(\dfrac{\alpha^{\pi-1} \delta^{3\pi} \gamma^{1.5\pi}}{\gamma^{0.5\pi} \beta^{2\pi} \alpha^{-1}} \right)^{1/\pi}$

16. What is the difference between discrete and continuous set of points?
17. What is the difference between an expression, an identity and an equation?
18. Is $a\%$ of b same as $b\%$ of a?
19. Explain the difference between inductive reasoning and deductive reasoning.
20. What is the meaning of locus of a point?

1 The Holy Laws of Mathematics and How to Break Them: Indeterminate Forms

> *"Mathematics is a place where you can do things which you can't do in the real world."*
>
> – Marcus du Sautoy

"Victory is our tradition" was the motto of one of the mightiest cruisers of the United States Navy – *USS Yorktown (CG-48)*. On September 21, 1997 – the ship suddenly had a massive system failure. Its entire network had crashed, the communication system halted and the ship's propulsion failed. For almost two and a half hours, the ship was in a "dead state". But what had caused this? The ship's computer system had attempted to break one of the most holy laws of mathematics –

Thou shalt not divide by zero.

The Laws of mathematics do not care about your traditions. There are things that are prohibited in mathematical scriptures and there are some unsaid rules which we all accepted that we grew up. We did all sorts of complicated problems in our tests – algebra containing scary decimal computations, unimaginable ways of finding roots, some stupid geometrical diagrams and what not! But never ever, we were even asked to divide something by zero. It is as if no one wanted to talk about it. That is somehow showed the weakness of mathematicians – and that they were embarrassed to talk about it. But let us try to understand, *what exactly is the problem in division by zero?*

1.1 Division by Zero

For most of the history of mankind, zero has remained a mere abstract concept. In poetry and philosophy, it represents "void", "space" or simply "nothingness". Often personally, zero is more than a number and has feelings attached to it – yet it has remained one of the most daunting things for mathematicians to deal with. If you happen to visit

Chaturbuj Temple in Gwalior (I guess it is dated around 870 A.D.), you will come across one of the oldest records of the symbol '0' engraved on the wall. The use of zero in the decimal system was first started by Aryabhatta. And while almost all big civilizations – Mayans, Sumerians etc. had the knowledge of the concept of zero – the Indians were the first to deal with it in the form of a number. It is quite difficult to imagine "nothingness" as a number – let alone divide by it. But in any case, we can try to get a "feel" of it. Look at the following observations:

$$\frac{1}{2} = 0.5,$$

$$\frac{1}{0.5} = 2,$$

$$\frac{1}{0.1} = 10,$$

$$\frac{1}{0.01} = 100,$$

$$\frac{1}{0.0001} = 10000,$$

$$\frac{1}{0.000000001} = 1000000000$$

And we can continue doing this, which loosely means:

$$\frac{1}{a \text{ very very small positive number}} = a \text{ very very large positive number}$$

Which by the way should also imply (I hope I do not need to show you this and you can test it for yourself):

$$\frac{1}{a \text{ very very small negative number}} = a \text{ very very large negative number}$$

But here is the real challenge: exactly *how small* of a number can we put in the denominator and exactly *how big* of a number we can get on the right hand side? This does not seem to have a definable answer to it. No matter how small of a number you can think of – I can still get you a smaller number and no matter how big of a number you think of – I can still get you a bigger number than that.

How does one resolve this? If we keep getting smaller and smaller, we reach a point where we have nothing left – denoted by the symbol '0'. However, one can get as big as one wants to. There seems to be no limit to it. Boundless. Coming here to our rescue is not Newton this time, but rather his editor of mathematical texts – *John Wallis*, who is credited with denoting "a very very large number" by '∞' (infinity). So, the above observation can be symbolically written as:

$$\frac{1}{0} = \infty$$

Let us be very much aware that we have not solved any problem – but rather we have merely represented it in the language of mathematical symbols. That's all. It still doesn't tell us "how big" of a number we will get as an answer. So that symbol ∞ is merely an idea – a concept – and **not a number**. It is quite evident that it is a huge number – but there is no definition as to how huge it is.

And so $\frac{1}{0} = \infty$ is one of the examples of what we call in mathematics as *"not defined"* case. We sure do have an image of the answer but we cannot *define* it. We should be happy at this point – at least we have an answer! But why am I saying so? Is there something fishy going on?

We discussed here one of another holy laws of mathematics that we must not treat infinity as a number. Because if it is a number – we must be able to apply the rules of algebra to it.

But what if we try? Let's see in illustration 1.1.

Illustration 1.1: Just for the sake of adventure, let us go ahead and do some algebra here:

$$\frac{1}{0} = \infty$$

$$\Rightarrow 1 = 0.\infty \ldots (i)$$
$$\Rightarrow 1 = (2.0).\infty \text{ [since } 2.0 = 0\text{]}$$
$$\Rightarrow 1 = 2.(0.\infty)$$
$$\Rightarrow 1 = 2.1 \text{ [}\textit{from (i)}\text{]}$$
$$\Rightarrow 1 = 2$$

Which is completely absurd! Did we just prove that 1 = 2? You might be thinking that why did we particularly choose 2 in the above working. Well, for that matter, we could have chosen any other number *a* on the planet and we would end up proving *1 = literally any number a on the planet*!

That is the complete breakdown of algebra right there. It is bizarre and makes absolutely no sense at all. In fact, according to illustration 1.1, you can go on proving that any number is equal to any other number. Now you know why I mentioned a few moments ago that you should be happy that at least you have an answer in the *not defined* case, because here you cannot even determine anything close to an answer here. It is an absolute mess. We call such cases in mathematics as the **indeterminate forms** – which is the central idea of this chapter (Notice that an algebraic operation with a not defined case has led to an indeterminate form).

Meme 1.1: Not allowed!

And since we are starting to get a little adventurous, let us dig deeper into this mess called the *indeterminate forms*.

1.2 Indeterminate Forms

Well, we just started to play and we figured out that $0 \cdot \infty$ is not something we can go about playing with. It also establishes this holy law that one must not consider ∞ as a number that he/she can play with but rather just an idea or a concept ("just a huge number"). Playing with $0 \cdot \infty$ resulted in quite some mess. There are seven such messy situations that we will frequently encounter in our study of Calculus. I list them down in the increasing order of difficulty:

1. $\dfrac{0}{0}$ 2. $\dfrac{\infty}{\infty}$ [not so messed up]

3. $0 \cdot \infty$ 4. $\infty - \infty$ [quite messed up]

5. 0^0 6. 1^∞ 7. ∞^0 [really messed up]

There could be more such indeterminate cases, but at least in my knowledge, if you know how to deal with these bad boys, you'll be good to go. We have already discussed the mess with $0 \cdot \infty$, let us look at the first one i.e. $\frac{0}{0}$ in the next illustration.

Illustration 1.2: Indian mathematician Brahmagupta was one of the first to deal with calculations involving zeroes. His achievements in mathematics and astronomy are countless. However, even a mathematical giant like him concluded that $\frac{0}{0} = 0$. Let us see the problem with this indeterminate form.

We know that

$$\frac{0}{any\ number} = 0 \ldots \text{(i), and}$$

$$\frac{any\ number}{0} = \infty\ (\textit{"a huge number"}) \ldots \text{(ii), and also}$$

$$\frac{any\ number}{any\ number} = 1 \ldots \text{(iii)}$$

Now, if that "any number" is 0 itself, then according to (i), (ii) and (iii), the answer can be either 1, 0 or ∞. It's quite messed up situation.

Meme 1.2: Logical talks!

Illustration 1.3: Let us see the problem with 0^0. We know from our experience of algebra:

$$(any\ number)^0 = 1 \ldots \text{(i), and}$$

$$0^{(any\ number)} = 0 \ldots \text{(ii)}$$

Again, the same question arises – what if that "any number" is 0 itself? According to (i) and (ii), the answer could be both 0 and 1. Which one do you prefer? And again, we end in an indeterminate situation where taking any one decision is quite difficult.

We can go on showing the problems with all the seven indeterminate forms, but I am too lazy and I think I have made my point here – *it is simply not "commonsensical" to work with them and in most cases our general rules and experiences of algebra breaks down.*

Query 1.1: Why is ∞^0 an indeterminate case but 0^∞ isn't?

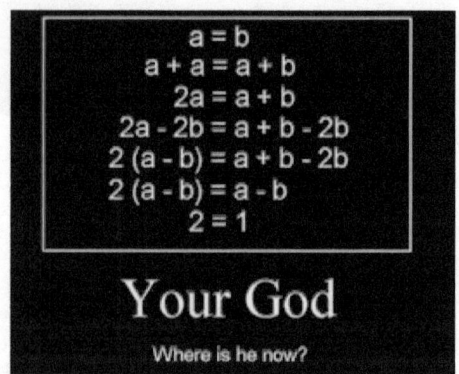

Meme 1.3: What is the problem here?

1.3 Closure

In the middle of the 19th century, when the Catholic priest Abbe Francois-Napolean-Marie Moigno, first coined the word *indeterminate terms*, he was probably just taking inspiration from the works of Cauchy. But are we talking about these messed up cases here? What relation do they have in the study of calculus? We saw how they gave messed up, and in most cases a nonsensical answer! Why can't we just ban them and move them. Well, here is the bad new folks:

The whole goal of Calculus would be to know how to deal with these indeterminate forms.

Meme 1.4: When the book asks you to prove something.

Exercises

1. A counterexample is an example that is used to disprove a mathematical statement. Here is a list of common mistakes that students do while studying mathematics. Use counterexamples to prove that they actually are mistakes (do not repeat them yourselves in the future):

   a. $(a+b)^2 = a^2 + b^2$

   b. $\sqrt{a^2 + b^2} = a + b$

   c. $\sqrt{a+b} = \sqrt{a} + \sqrt{b}$

   d. $a^m \cdot a^n = a^{mn}$

   e. $\dfrac{1}{a} + \dfrac{1}{b} = \dfrac{1}{a+b}$

   f. $\dfrac{a^m}{a^n} = a^{m/n}$

   g. $\dfrac{A}{B+C} = \dfrac{A}{B} + \dfrac{A}{C}$

2. What is meant by a *fallacy*?

3. What is more correct to say: *Parallel lines never intersect* or *Parallel lines intersect at infinity*?

4. For what value of x, does the expression $\dfrac{x^2 - 4}{x - 2}$ become indeterminate?

5. Research more about Riemann's paradox which says $\pi - \pi = \infty$.

2 Our Place in the Universe and Mathematics: Coordinate Systems

"Obvious is the most dangerous word in mathematics."

– E. T. Bell

There is a famous legend in physics about Sir Issac Newton discovering the law of gravitation as the apple fell on his head while he was sitting below the apple tree. There is an equally fascinating (may be not so famous) legend in mathematics about the person who paved the way for most of Newton's mathematical work – Rene Descartes, who was a 17th century French mathematician and philosopher. Due to his fragile health conditions, he was mostly advised bed-rest. Laying on his bed, he used to wonder about mathematics, philosophy, life and about existence. It was on one of these days of sickness as he was lying on his bed and was staring up at the ceiling. A fly landed upon the ceiling and his eyes started to follow it. He asked himself a simple question, *"How do I tell the position of this fly roaming about on the ceiling to someone else?"*. There in that moment, he had a series of revelations – all following each other. Each thing leading to another and he had finally achieved what would revolutionize mathematics for ages to come. But, how?

This is the story of Descartes, his fly and all of modern mathematics.

2.1 Descartes and His Fly: The Cartesian Plane

Let us get back to the sick Descartes staring at the fly on his ceiling, could be something like this in Fig 2.1 as shown.

Fig 2.1: The fly on ceiling

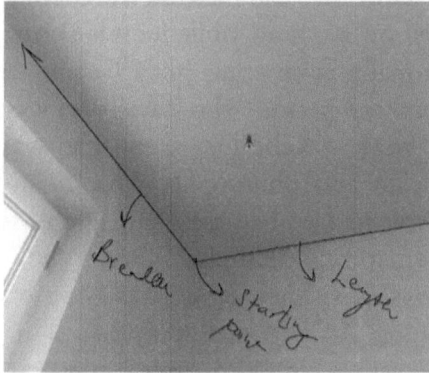

Fig 2.2: The fly on the ceiling

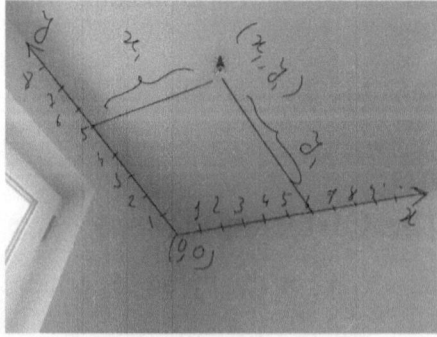

Fig 2.3: The fly and birth of coordinates

In his head, he imagined the mutually perpendicular length and breadth of the ceiling as *reference lines* and one corner to be the *reference point* (like in Fig 2.2).

He chose to call the length as the *x-axis* and the breadth as *y-axis* (just a mathematician's habit of naming words with symbols/letters). Then, he could mentally divide the x-axis/length and the y-axis/breadth into smaller divisions of the same scale. We can then start allocating the numbers considering the starting point as the corner as decided. If the fly happens to be x_1 distance away from starting point or *origin* in length and y_1 distance away from starting point or *origin* in breadth (Well, x_1 and y_1 could be any numbers depending upon where fly is at that moment, which means of course, they are *variables*), Descartes suggested that one could represent the position of the fly as (x_1, y_1) (You can always imagine (x_1, y_1) in your head as (length from origin, breadth from origin), simple!), as shown in Fig 2.3:

I guess, it is obvious that the corner/origin must be represented by (0,0).

Now, why he chose to write length first and breadth second

is something only he knows – but we just follow the convention i.e. we will **always** write length first and breadth second. The whole ceiling could be called a *Cartesian Plane* (rightfully named after him).

But then, what is the big deal? How far this simple thought even lead to? Before knowing the consequences, we must first make things a bit more formal (for our future use, without referring to the fly and the ceiling each time.)

2.1.1 Putting Things a Bit More Formally

Now, we will not always be dealing with the fly and the ceiling – so one must generalize the system Descartes came up with. Refer to the diagram in Fig 2.4:

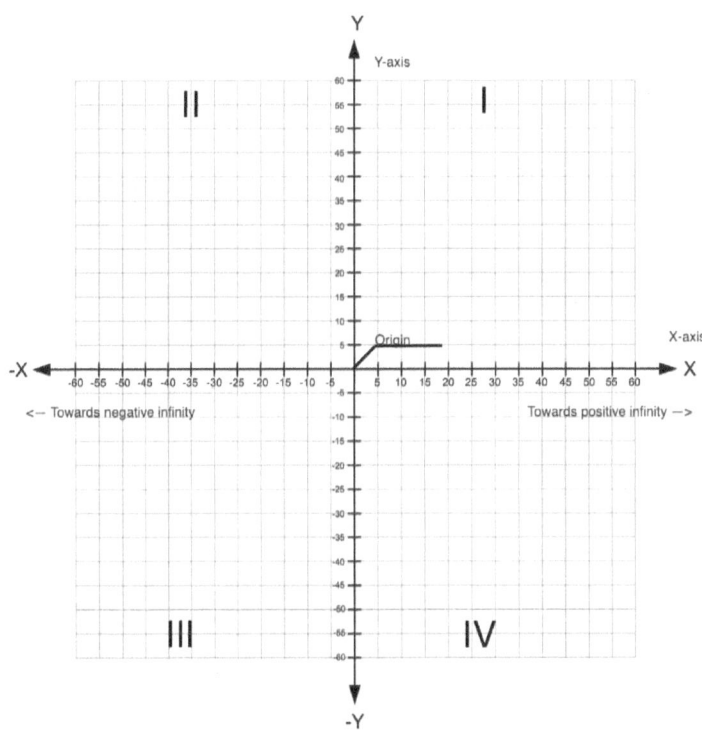

Fig 2.4 Cartesian Plane and coordinate system

We are free to choose the starting point of our measurement. That point will be called *origin O* having the position value as (0,0). We now lay down a horizontal line through O and call it the x-axis. We can mark off measurements at equal distances. Towards the right side of origin, we mark positive numbers and towards the left, we mark off negative numbers (yet another convention). Through we now draw a line perpendicular to the x-axis through origin and we call it as the y-axis. Going up the origin, we mark positive numbers at equal distances and going downwards the origin we mark the negative numbers at equal distances. The entire plane containing the x and y axes is called *Cartesian Plane*. Note that the x – and the y – axes are *infinitely long*.

Notice that due to the axes, the entire Cartesian plane has been divided into four parts – we call each part as a *quadrant*. Let us go into a little bit more detail into this. Look at the following four different positions on the cartesian plane marked by A, B, C and D in the following diagram.

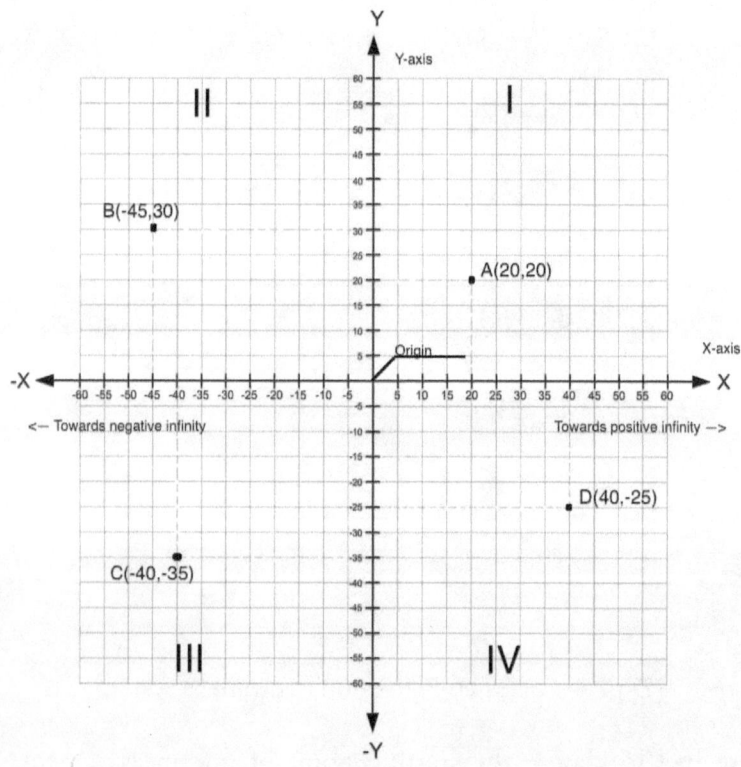

Fig 2.5: Cartesian Plane and coordinate system

Table 2.1: Mapping points to the cartesian plane

A(20,20)	To reach the location A, you must move 20 units right (+) from the origin and 20 units (+) above the origin. This area in which we always have to move right of origin/+ve value of x and upwards from origin/+ve value of y is called the 1st quadrant.
B(−45,30)	To reach the location B, you must move 45 units left (−) from the origin and 30 units (+) above the origin. This area in which we always have to move left of origin/−ve value of x and upwards from origin/+ve value of y is called the 2nd quadrant.
C(−40,−35)	To reach the location C, you must move 40 units left (−) from the origin and 35 units (−) below the origin. This area in which we always have to move left of origin/−ve value of x and downwards from origin/−ve value of y is called the 3rd quadrant.
D(40,−25)	To reach the location D, you must move 40 units right (+) from the origin and 25 units (−) below the origin. This area in which we always have to move right of origin/+ve value of x and downwards from origin/−ve value of y is called the 3rd quadrant.

Any location (x, y) on the Cartesian Plane is called an *ordered pair* (as the order matters here). Thus, just by looking at the symbol of the numbers in the ordered pair, we can tell which quadrant it lies in. This has been summed up in the following table:

Table 2.2: Signs of points in quadrants

(+,+)	1st Quadrant
(−,+)	2nd Quadrant
(−,−)	3rd Quadrant
(+,−)	4th Quadrant

2.1.2 *Knowing How Far We Are: The Distance Formula*

For starters, I must tell you that this is nothing but literally just Pythagoras' Theorem in a different manner. Let us take two random points on the Cartesian plane, P($x1$, $y1$) and Q($x2$, $y2$). Our goal is to

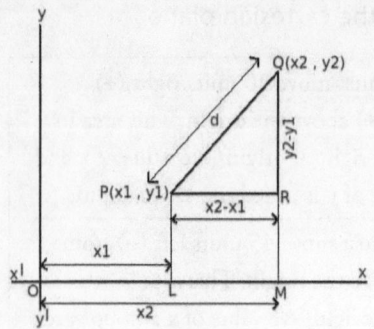

Fig 2.6: Derivation of distance formula

know what is the length of the straight line joining them (or more simply, what is the distance between them?). Let us refer to Fig. 2.6.

If we drop a perpendicular from P to x-axis and name it L, we can clearly see that in order to reach to P, we must walk OL = x_1 units on the right and LP = y_1 units upwards. The same could be done with Q, let's drop a perpendicular from Q to x-axis and we name it M, we can clearly see that in order to reach to Q, we must walk OM = x_2 units on the right and MP = y_2 units upwards. Also, notice that clearly MQ and LP are parallel to each other. Finally we drop a perpendicular from P to QM at R, so that PQR forms a nice right-angled triangle. Also notice that PR and LM are parallel and equal to each other now, the same holds for PL and RM. *(Why?)*

Now, PR = LM = (OM − OL) = $x_2 − x_1$ units, And, QR = (QM − RM) = $y_2 − y_1$ units

Finally, by Pythagoras' Theorem:

$$PQ^2 = PR^2 + QR^2$$
$$\Rightarrow PQ = \sqrt{PR^2 + QR^2}$$
$$\text{or, } d = \sqrt{(x_2 − x_1)^2 + (y_2 − y_1)^2}$$

And there we go!

2.2 Why I Call Descartes the "Buddha" of Math: Analytical Geometry

(I would advise you to go slow with this section. It is something which will stick with you for a long long long time in the due course of this book and later practices in Vol. 2.)

2.2.1 Geometry

What comes to your mind when you hear the word *"geometry"*? May be a bunch of 2-D and 3-D figures like shown in Fig 2.7 or just a bunch of formulas for area and what not as shown in Fig 2.8.

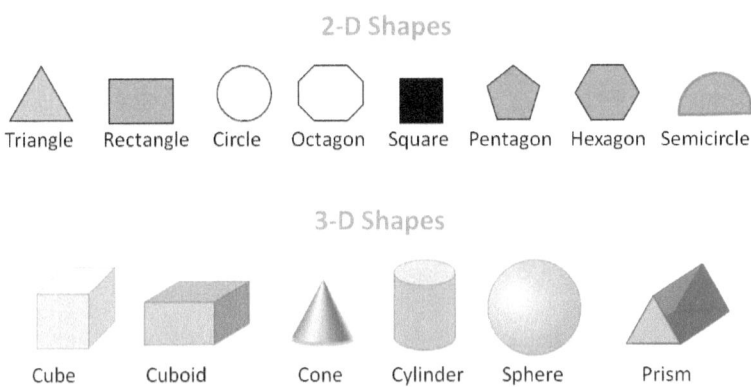

Fig 2.7: Bunch of 2-D and 3-D shapes

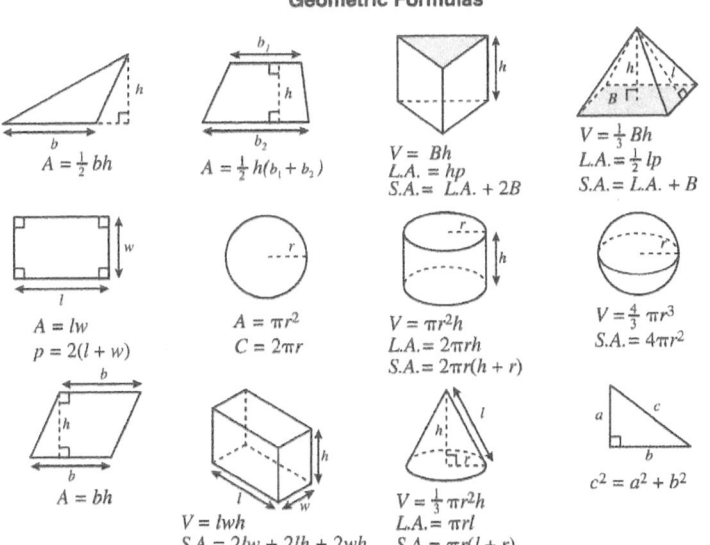

Fig. 2.8: Bunch of geometrical formulas

I assume whatever might come to your head, one thing is obvious: that geometry is everywhere – the shapes, the patterns, the recurring sequences, the periodic motions of planets etc. And also, that you have to *draw* it in order to study it. I mean there is always a pencil, paper, compass, protractor, angles and lines associated with it. Well, good enough. Let's see the next big thing.

2.2.2 Algebra

Growing up, personally I have always **hated** algebra. I mean it. The reason lies in its *abstractness*. It says a lot of things but there is no *image* to it. In fact, algebra is one point from which most of the learners sort of loose math (just a personal opinion). In fact, what comes to your mind when you hear the word *"algebra"*?

May be some crazy equations that we had to solve to get past the exams, like:

$$x^2 + \sqrt{3}x + 7 = 0 \text{ or } y = ax^2 + bx + c$$

or the boring long long polynomial divisions like the one in Fig 2.9:

$$
\begin{array}{r}
x^2 + x + 3 \\
x-3{\overline{\smash{\big)}\,x^3 - 2x^2 + 0x - 4}} \\
\underline{x^3 - 3x^2} \\
+x^2 + 0x \\
\underline{+x^2 - 3x} \\
+3x - 4 \\
\underline{+3x - 9} \\
+5
\end{array}
$$

Fig. 2.9: Eh!

or probably the word problems that made no sense to anyone:

"Gabby is 1 year more than twice Larry's age. 3 years from now, Megan will be 27 less than twice Gabby's age. 4 years ago, Megan was 1 year less than 3 times Larry's age. How old will Megan be 3 years from now?" (like honestly, WHO CARES?).

And with all those big formulas, identities and equations on the blackboard in front of us, at the end of the day, we wonder: *where will I even use it in real life?*

2.2.3 Descartes, Geometry, Algebra and the Enlightenment

Let us go back to the crazy idea that Descartes had while watching the fly on the ceiling. Every ordered pair like (x, y) represents a point. By now, you know how to locate it on the Cartesian plane.

And all shapes are made of so many points, right? Here is where the magic happens now. Think of a boring algebraic equation. Let me write the one that comes to mind now:

$$y = x^2$$

Our trained mind gives us boring answer related to this equation: It's a second-degree polynomial equation blah! blah! But let the numbers work now. Let's plug in some values of x here – and we get one value of y, don't we? What values of x can we put? Any. Infinite. As much as we want. Let's try:

x	-5.5	-5	-4.5	-4	-3.5	-3	-2.5	-2	-1.5	-1	-0.5	0	0.5	1	1.5	2	2.5	3	3.5	4	4.5	5	5.5
y	30.25	25	20.25	16	12.25	9	6.25	4	2.25	1	0.25	0	0.25	1	2.25	4	6.25	9	12.25	16	20.25	25	30.25

Fig. 2.10: Some values for $y = x^2$

But all these ordered pairs like (–5.5, 30.25), (–5, 25), (–4.5, 20.25) ... (4.5, 20.25), (5, 25), (5.5, 30.25) represent so many points on the plane. What happens when we draw all these points? When we try that, we get a figure like shown in Fig. 2.11:

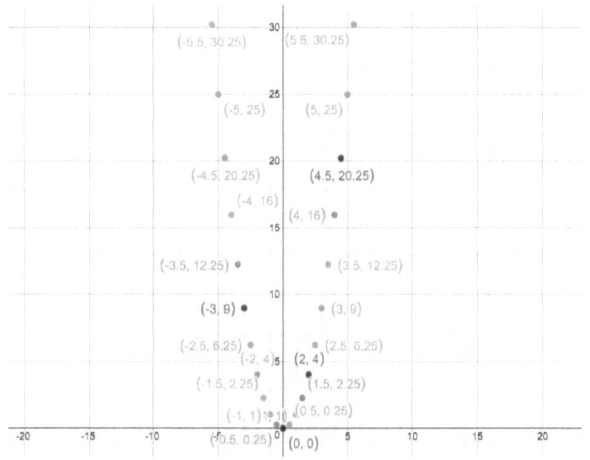

Fig. 2.11: Putting some values for $y = x^2$

Joining all these little points (and all those infinite points in between) by a smooth curve, we get something as shown in Fig. 2.12:

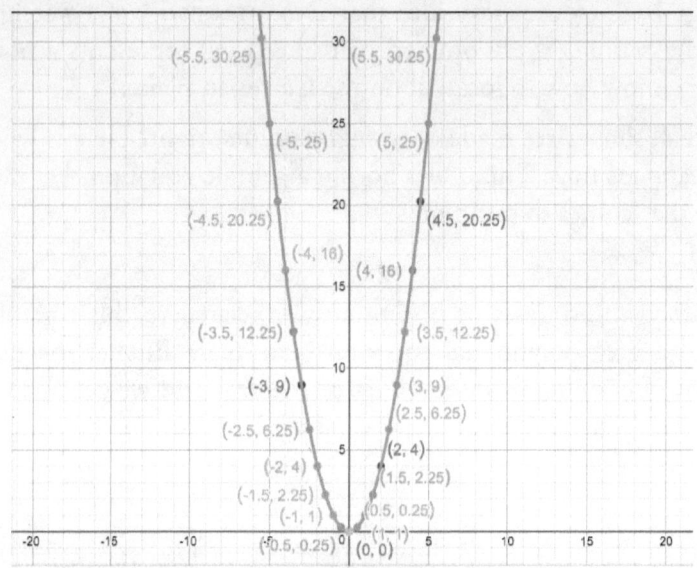

Fig. 2.12: A nice, neat curve for $y = x^2$

And there! The boring algebraic equation $y = x^2$ suddenly seems to have started to speak. It is kind of like an "instruction" for the points to *"behave"* in a certain manner. Suddenly, there is a strange, uncomfortable revelation here that these gibberish symbols like $x^2 + y^2 = 1$, $y = x^3$, $y = x$ all have *"hidden pictures"* in them. They are trying to talk to us. They are painting a picture, a shape and a geometry – a certain type of behavior of the selected ones of those infinite points in the Cartesian Plane.

The rules are quite simple: Take any random algebraic equation, keep plugging values of x, get some y as answer and there you have a point! Get a good number of these points, join them and suddenly you get a picture. They are not just x and y anymore. I wonder that when this idea would have struck Descartes – he would have sat on his table in awe of what these symbols could do now. The geometry existed in the coded algebra and suddenly nothing is the same anymore.

All the algebraic equations you ever came across have something to tell you – they are waiting to be painted on the cartesian plane. How many of them? Infinite. Infinite equations. Infinite shapes. Infinite possibilities.

The nature – the vast and elegant nature that is so geometrical manifests itself as a form of algebra too. Who would have thought?

All because, one night, one man wondered looking up – *"How do I tell the position of this fly roaming about on the ceiling to someone else?"*

Descartes bridged this gap between algebra and geometry into what we call these days as *analytical geometry*.

Fig. 2.13: The Enlightenment

Next few chapters of our book will be dedicated to a deeper understanding of this art that we have just been introduced to. Here, you are just beginning to learn the language of the nature – the language of mathematics. Sit back and relax and let your mind picturize every single equation that you see hereafter. We shall continue this section again in Chapter 4. Let us look at some other ideas now for a change.

Illustration 2.1: Algebraic or geometric method? For many questions in mathematics, we have the choice to go about them in any way feel we like. Let me give you one example here. If we take two numbers a and b. We have something called Arithmetic Mean (which is $\frac{a+b}{2}$) and Geometric Mean ($\sqrt{ab}$) (of course, you don't have to understand them at this point). There is a statement about them which says:

$$A.M. \geq G.M \text{ or } \frac{a+b}{2} \geq \sqrt{ab}$$

We have to prove that this statement is indeed true. Let's go for the algebraic method first.

We know that the square of any number is always greater than or equal to zero. Thus, we can say,

$$(a-b)^2 \geq 0,$$

or, $a^2 - 2ab + b^2 \geq 0, \Rightarrow a^2 + b^2 \geq 2ab$

$\Rightarrow a^2 + 2ab + b^2 \geq 4ab$ (Adding 2ab both the sides)

$\Rightarrow (a+b)^2 \geq 4ab, \Rightarrow (a+b) \geq 2\sqrt{ab}$ (Taking roots of both the sides) and finally, $\dfrac{(a+b)}{2} \geq \sqrt{ab}$,

Basically, $A.M. \geq G.M$

Now, this is not a very difficult proof to follow, but I must admit that it is pretty random. For example, why did I start with $(a-b)^2 \geq 0$. Could have started with anything else. So it is a serious of trained and experienced guess in algebra that will lead us to the desired answer. All having said, there is no picture to it as such to the whole problem – it doesn't look obvious or you know, intuitive – like you just know it.

Let's try the geometric picture next. We imagine these two given numbers a and b are length of some line segments. Therefore, AB measures a unit in length and BC measures b units in length. Now, taking AC (which has a length of $a + b$) as diameter, we make a semi-circle (F being the center and radius being $\dfrac{a+b}{2}$). The line DF (which we draw perpendicular to AC) therefore also measures $\dfrac{a+b}{2}$ (DF also being the radius). Now, from point B, we draw a perpendicular to a point E on the circle. With little bit of geometry and properties of circles, we can find that the length of EB is actually $\sqrt{ab}$ (try to verify this) as shown in Fig. 2.14.

Therefore, FD represents the A.M. and E.B. is the G.M. Here is the important part now, No matter what the values of a and b, DF will

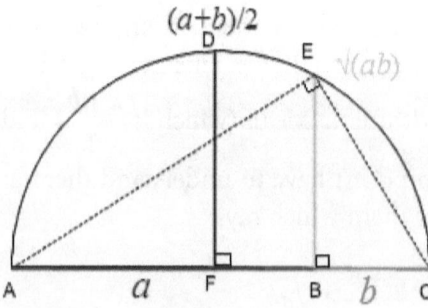

Fig. 2.14: Geometric representation of A.M. and G.M.

always remain just mid-way of the semi-circle, however, *EB* will change its position depending upon the values of *a* and *b*. But since, radius is the maximum length that is allowed in a semi-circle – *EB* can at max. be equal to *DF* or smaller than that – never greater! In other words, $A.M. \geq G.M.$ And that's all! We not only have a proof of the statement, but also can "see" the proof!

Personally, I find geometric method more interesting and intuitive – it gives me vision; however, algebra lets me feel.

2.3 Why Would Newton Stay Behind: The Polar Coordinate System

Here is the thing about Newton: He will sneak in everywhere! Mathematics, physics, philosophy and what not! Decades after the Cartesian system of representing points/locations, while reading them – Newton came up with TEN NEW coordinate systems! Ten new! One of them he called as the polar co-ordinates. Although I must admit that at this point it is quite unnecessary to introduce polar coordinates as you would not use them immediately – I am writing it here just because of my bias. Once you start playing with polar coordinates – you get wonderful shapes! Trust me! Most of them would be a nightmare to achieve by the Cartesian Plane and equations. Anyway, you may skip this section and come back to this when you encounter it for the second time in the next book!

Meme 2.1: *NEWTON*

2.3.1 Polar Coordinates

Here is the basic idea: Let us look at a point A(1,1) in the cartesian plane as shown below:

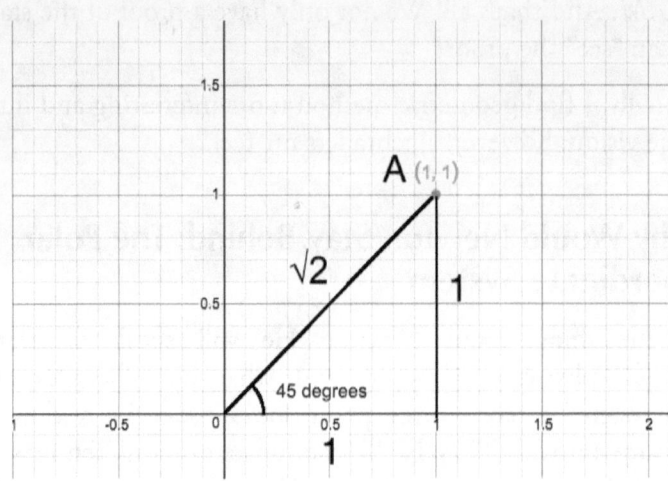

Fig. 2.15: (1,1) on Cartesian Plane

By distance formula discussed before, it is quite evident that the distance of (1,1) from origin (0,0) is simply $\sqrt{(1-0)^2 + (1-0)^2} = \sqrt{2}$. The line OA makes 45° with the x-axis which is shown in Fig. 2.15. Which basically means that if you take a line of length $\sqrt{2}$ on the x-axis and rotate it 45° in the anti-clockwise direction – you still reach the point A(1,1).

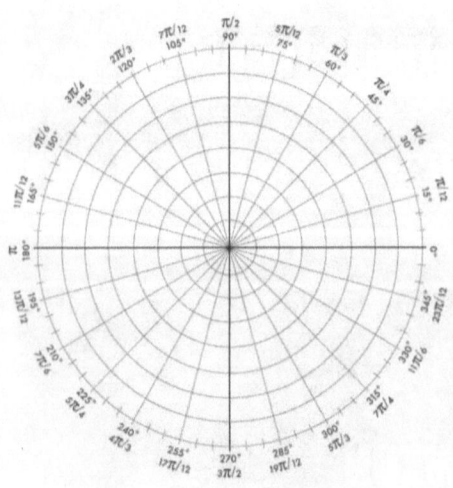

Fig. 2.16: The Polar Plan

Thus, Newton figured out that the same point can be reached provided you the following information: a) How far it is from origin (denoted by *r*) b) How much angle it makes in anti-clockwise direction from x-axis (denoted by ϑ). A typical Polar plane looks like its shown in Fig. 2.16.

Since, for a given length we can move the entire 360 degrees, thus it contains circles of different radii. Here the coordinates will have the form (r, ϑ).

Query 2.1: How do you interchange mathematically between cartesian and polar coordinates?

Exemplum 2.1: Draw the graph of Polar equation $r = 1.25^\theta$

Solution: The idea again is quite simple here: you chose an angle ϑ, you get a length r and you mark a few points. A table of values is shown in Table 2.3.

Table 2.3: Set of values for $r = 1.25^\theta$

θ	0°	45°	90°	180°	360°	540°	720°
$r = 1.25^\theta$	1	1.1915	1.4197	2.0158	4.0635	8.1913	16.5122

We see that with increasing angles, the corresponding radius also increases and therefore, the graph is bound to be a spiral as shown in Fig. 2.17

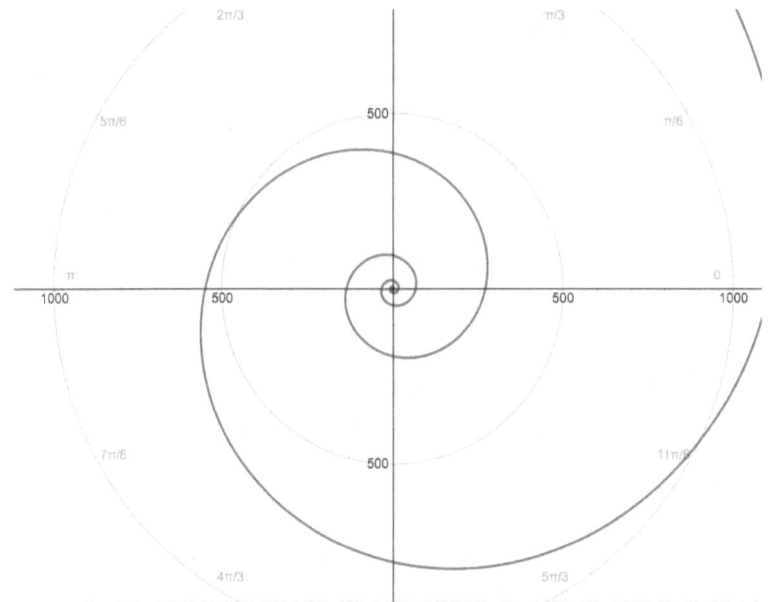

Fig. 2.17: Polar graph for $r = 1.25^\theta$

The beauty of polar graphs is that they seamlessly and quite effortlessly give you beautiful curves. Anyway, we shall come back to this in Vol.2 of this book perhaps. For now, you are free to make equations including r and ϑ and look out for more beautiful graphs with them.

2.4 Degree vs. Radian Measure

Ever wondered why the total angle of a circle measures 360°? I mean it seems pretty random to divide the circle in 360 parts. Of course, it would look natural to you because that's what you've been seeing since childhood – but question it now. Why 360, why not any other nice number 400, or 1000 – I mean literally anything.

Unfortunately, there is no definite answer to this – but there are possibilities. The ancient Sumerians, Babylonians and Indians used the sky to make calendars (apart from the fact being that they used sexagesimal system). It appeared to them that the sun passes through a set of 12 zodiac constellations in the sky and took about 360 days to come back to the original constellation it started out with as shown in Fig. 2.18

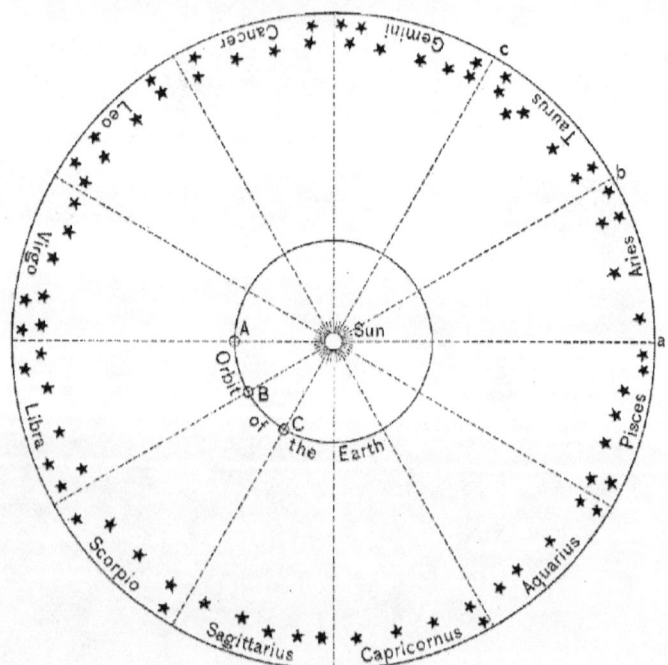

Fig 2.18: Motion of sun through the zodiac constellations

This does have obvious relations to zodiac signs and astrology (but if by any chance you believe in astrology, I must remind you that it's the 21st century and it's good to come back here). So, they divided the sky (they considered the path of the sun to be circular) into 360 parts. Another strong reason being that the number 360 has got a good number of divisors – 24 to be precise. Meaning, you can divide 360 into 24 smaller parts without having to deal with nasty decimals. Seems like a fair deal!

But as you go further in mathematics and science, you will notice another unit of measuring angle called the radian measure. Now, if we were already happy with having the sexagesimal or the degree measure – why bother making a new angle measurement system?

2.4.1 The Radian Measure

The radian measure was defined by James Thomson in 1873 (who was elder brother of the famous physicist Lord Kelvin – the guy who was too sure that we have discovered everything that is there to discover in physics. Within 5 years of this statement, Einstein comes and rocks the entire scientific world with a whole new type of physics).

The radian (θ) is defined as the ratio of the length of the arc (s) of a circle to its radius (r). Symbolically, $\theta = \dfrac{s}{r}$.

If we take radius as 1 unit then we have, $\theta = s$. That means in a circle of radius 1 unit (also called the unit circle, to which we will come back in Chapter 7), the radian angle measure is same as the length of the arc of the circle. The circumference of such a circle would be $2\pi(1)$ units or simply 2π (which is the equivalent radian measure). But a complete circle subtends an angle of 360°. Therefore, we have the equivalence:

$360° = 2\pi$ *radians*, or

$1 \, radian = \left(\dfrac{180}{\pi}\right)°$

If the arc length is equal to the radius, then we can write:

$\theta = \dfrac{r}{r} = 1 \, radian$

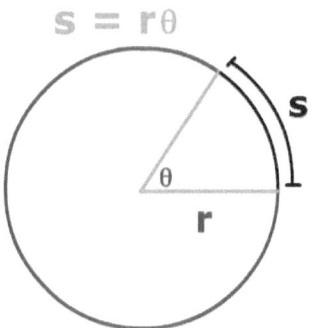

Fig. 2.19: Radian Measure

But these are just technicalities, where is radian actually important? And why do we even use it. I would prefer to keep you in suspense for now and come back to this when we actually start doing Calculus. Till then, you are free to look around and get used to it. *(A small note here: Angles are always measured in anti-clockwise direction from the x-axis. Clockwise measurement of angles are taken in negative units).*

Query 2.2: Convert the following angles into radian measure a) 45° b) 180° c) 270° d) 60° e) 30°

2.5 Let Me Tell You a Magical Story

We discussed quite a few concepts in this chapter. One of the most important was the concept of *knowing our position via a reference point*. In fact, it is one of the oldest philosophical questions in the history of Mankind. Where are we in time? And where are we heading to? To be honest, these concepts of reference points and coordinates look so abstract. They just don't seem to serve any real purpose.

Or, do they?

On April 11, 1970, three NASA astronauts James A. Lovell, John L. Swigert and Fred W. Haise Jr. launched off from the surface of earth aboard Apollo 13. They were to walk on the surface of the moon two days later. But 56 hours into the mission and 330,000 km away from Home Earth – an accident occurred. Houston flight controllers asked Swigert to turn on the hydrogen and oxygen tank stirring fans in the Service Module, which were designed to de-stratify the cryogenic contents and increase the accuracy of their quantity readings. Two minutes later, the astronauts heard a "pretty large bang", accompanied by fluctuations in electrical power and the firing of the attitude control thrusters.

So there were three people – floating in the dead empty space on a damaged aircraft with petty quantities of oxygen, collection of increasing amounts of toxic carbon dioxide in the escape ship and every single thing out there to kill them.

Yet they had to come home. This was a challenge NASA never had before. Next few days, after hundreds of engineers and scientists thrusting their collective knowledge and work to get them back somehow – they were close to home.

Except for one small problem.

In order to navigate back to earth, they needed instructions for direction that would be fed on the computers. But then, there was no power to switch on the computer. They needed one point of fix the altitude and then fly the spaceship. But due to the damage, there were a lot of debris and the stars were not visible. There was possibly no hope.

Until James A. Lovell saw something through the window of the shuttle:

The Earth

Keeping Earth as the reference point, Swigert flew straight onto it – thus making it one of the most successful missions in the history of human space endeavors. The concept of reference points and frames was once developed on Earth; that day Earth acted as a reference point to save three of our bravest explorers.

Exemplum 2.2: A point P is equidistant from R(–6, 7) and S(3, –7) and its x-coordinate is twice its y-coordinate. Find the point P.

Solution: Since, distance are involved, we can be sure of using the distance formula here. Let the coordinates of the unknown point be P(2y, y) (since it's already given that x-coordinate is double of y-coordinate. The distance of R(–6, 7) from P(2y, y) can be calculated as:

$$RP = \sqrt{(2y-(-6))^2 + (y-7)^2} \quad ...(i)$$

Similarly, the distance of S(3,–7) from P(2y, y) can be calculated as:

$$SP = \sqrt{(2y-3)^2 + (y-(-7))^2} \quad ...(ii)$$

Since, both RP and SP are of equal distances, from (i) and (ii), we get:

$$\sqrt{(2y-(-6))^2 + (y-7)^2} = \sqrt{(2y-3)^2 + (y-(-7))^2}$$

$\Rightarrow (2y+6)^2 + (y-7)^2 = (2y-3)^2 + (y+7)^2$ (squaring both sides)

$\Rightarrow 4y^2 + 24y + 36 + y^2 - 14y + 49 = 4y^2 - 12y + 9 + y^2 + 14y + 49$

$\Rightarrow 10y + 36 = 2y + 9$

$\Rightarrow y = -27/8$

And therefore, $2y = -27/4$. Thus the coordinates of the point P are $(-27/4, -27/8)$.

Exemplum 2.3: Check if the three points A (2, 4), B (4, 6) and C (6, 8) are collinear.

Solution: Collinear points are the points that lie on the same line. Since, there are three points given – we can have the lines – *AB, BC and AC*. Two of these lines should total up to the length of the third line (Why?). Now,

$$AB = \sqrt{(4-2)^2 + (6-4)^2} = \sqrt{4+4} = 2\sqrt{2}$$

$$BC = \sqrt{(6-4)^2 + (8-6)^2} = \sqrt{4+4} = 2\sqrt{2}$$

$$AC = \sqrt{(6-2)^2 + (8-4)^2} = \sqrt{16+16} = 4\sqrt{2}$$

Now, clearly AB + BC = AC. Hence, the points are collinear.

2.6 Closure

This chapter is sort of a gateway to Chapter 4 about functions. Obviously, there are a number of topics that have been missed out in the formal sense like different forms of equations of straight line, section formula, midpoint formula etc. But the core idea is in the concept itself. I advise you to play with them as much as you can. Try extending the idea of two-dimensional coordinate system to three dimensions. A few other forms of coordinate systems like the spherical coordinate system and cylindrical coordinate systems shall be introduced when it'll required. However, that should not stop you from reading about them. Getting comfortable with notations is the prime step here.

Exercises

1. If you rotate the point A(2,3) by 180°, what are the coordinates of the new point? What about rotation by 90° and 540°?

2. A triangle is formed by joining the points A(3,4), B(6,2) and C(1,−4). Draw the triangle on the graph paper. Consider the x-axis and the y-axis as mirrors and reflect the triangles about both of them to make new triangle. What are the coordinates of the two new triangles formed? Do you see any pattern? If yes, what?

3. There are two given points, namely A(3,4) and B(−1,6). Find the coordinates of the third point C such that A, B and C form an equilateral triangle.

4. Convert the following to degree from radian measure:

   a. $\dfrac{11\pi}{7}$

   b. $\dfrac{3\pi}{4}$

   c. 360

   d. $\dfrac{-11\pi}{7}$

5. Convert the following to radian from degree measure:

   a. 720°

   b. −279.18°

   c. 57°30"

   d. $\pi°$

6. We know that in general $(a,b) \neq (b,a)$. Discuss all such cases where indeed $(a,b) = (b,a)$. Is there a particular equation that contains all of these points? If yes, what?

7. Ahmedabad is located at 23.0225° N, 72.5714° E and Guwahati at 26.1445° N, 91.7362° E. Considering earth as a perfect sphere of radius 6400 km. Calculate the curvilinear distance between the two places. (Hint: You only need to use the longitude measurement here)

8. Consider a circle with angle of sector θ. Find the formula for area of section (assuming you know the formula for the area of the circle) if θ is measured:

   a. Radians

   b. Degrees

   c. an imaginary unit where one right angle is 200°

9. A triangle is made up by joining three points $A(x_1,y_1)$, $B(x_2,y_2)$ and $C(x_3,y_3)$. Find the area of this triangle ABC in terms of these coordinates.

10. Can the formula of area of triangle obtained in problem 9 be used for proving collinearity of the three points?

11. Piyush has shifted to a new city and his parents are looking for a new school. They want a school that is within 5 km of distance from their home. If their home has coordinates of H(2,3) with respect to the centre of the city being C(0,0), write the algebraic equation of all regions within which the school must lie. Graph the equation and tell its shape.

12. The coordinate system is based on human sense of right-left-up-down directions. But suppose you are talking to an alien civilization that do not happen to use the same coordinate systems as us, how will you tell them the idea of directions? Is there any universal basis? You are only allowed to speak to them and not send photos or videos.

13. Let there be two numbers a and b. The Root Mean Square (R.M.S.) of these numbers is defined as $\sqrt{\dfrac{a^2+b^2}{2}}$, the Arithmetic Mean (A.M.) is defined by $\dfrac{a+b}{2}$, the Geometric Mean (G.M.) is defined by $\sqrt{ab}$ and the Harmonic Mean (H.M.) is defined by $\dfrac{2}{\dfrac{1}{a}+\dfrac{1}{b}}$. With reference to Illustration 2.1, prove the following inequality algebraically and geometrically:

    $R.M.S \geq A.M. \geq G.M. \geq H.M.$

3 Let's (not) Blame Euler for This: 0, 1, e, i, π

यथा शिखा मयूराणां, नागानां मणयो यथा।
तद् वद् वेदाङ्गशास्त्राणां गणितं मूर्धनि स्थितम्॥

"Like the crest of the peacock, like a gem on the head of a snake, so is mathematics at the head of all knowledge."

Now that we are slowly getting into the basics of precalculus, before getting any further – I would like you to get comfortable with five mathematical constants: *0, 1, e, i, π*. They look easy and harmless and there is a good chance that you have seen most of them before. I say *most* of them. But a little closer look – and you will slowly find out that they are not as plain and simple as they look. In fact, they contain deep mysteries within them. But worry not! We shall tackle them one by one. I expect that you are familiar with 0 and 1 at this level, so I begin with the *number e*. But before I even begin talking about *e*, here is something I must tell you. Let's go!

3.1 The constant e

This section is about how the number *e* came into being. A lot of books use *e* frequently but seldom talk about how interesting this number is.

3.1.1 Oh Dear, Bernoulli(s)!

As a young child interested in space science, I often wondered why things fly? Why are airplanes shaped the way they are? What is so special in birds? It turned out that anything that flies must obey this principle called *Bernoulli's Principle* (which obviously I will not discuss now). It intrigued me. In fact, the more I studied higher mathematics and physics – there were a number of things (and all interesting) named after this guy Bernoulli! For example, Bernoulli Triangle, Bernoulli Beam equation, Bernoulli distribution, Bernoulli differential equation and what not! I thought this guy must have been one hell of a genius. He was everywhere! Except there was one problem

This was not one guy but a full family.

Well yeah, the Bernoullis were a family of geniuses who produced about a dozen brilliant academics who made major contributions to mathematics and physics – and in a way shaped most of the modern mathematics that we study today. In fact, here is a family tree of Bernoullis shown in Fig. 3.1.

Out of these, three of them outshine the rest in the history: Jacob and Johann Bernoulli (who were brothers) and Daniel Bernoulli, who was the son of Johann Bernoulli. Probably one of the rare times when mathematics was a family business. But why do I mention them? Because our search for the constant *e* takes us back to Jacob Bernoulli and the problem that he was interested in. And I did not just draw the whole family chart to tell you only this, of course, there is more to it.

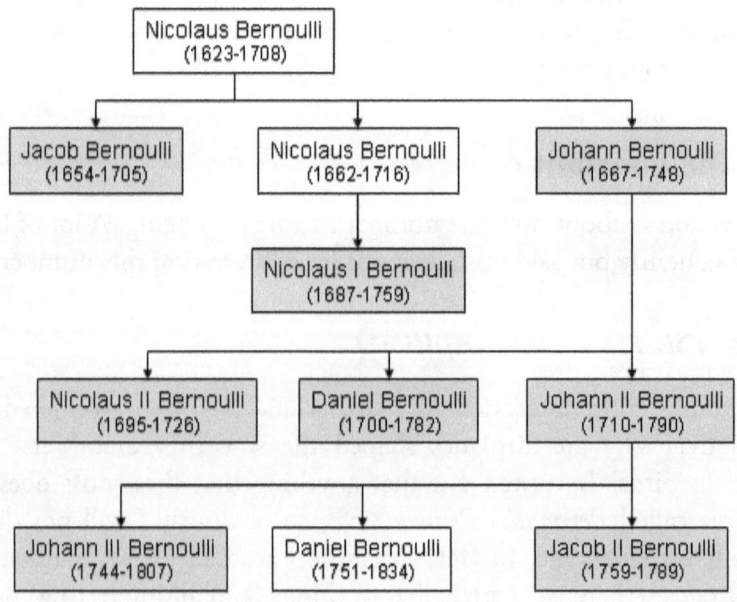

Fig. 3.1: The Bernoulli family tree

3.1.2 Jacob Bernoulli and Infinite Money

So Mr. Jacob was working on the problem of compound interest (the chapter I so hated in my school days). Let me break down the problem for you. Suppose you start with Re. 1 in your bank account. And then, hypothetically, you have managed to find a bank which gives you 100% interest compounded every year (that's too kind of them, but then who is stopping us from dreaming such things?). Therefore, at the end of 1 year, you have:

$$Re.\,1 + 100\%\ of\ Re.\,1 = Rs.\,(1+1) = Rs.\,2$$

Which seems good! So you get Rs. 2 at the end of the year. He then asked the following question: *Is it any better if I ask the bank to compound my money by 50% every 6 months?* He bought the interest by half, but then halved the time too. By common sense, it should not make much of a difference.

Let us see what happens.

At the end of the first six months, you get:

$$Re.\,1 + 50\%\ of\ Re.\,1 = Rs.\,(1+0.5) = Rs.\,1.5$$

And at the end of the next six months,

$$Rs.\,1.5 + 50\%\ of\ Rs.\,1.5 = Rs.\,(1.5+0.75) = Rs.\,2.25$$

Which means, you actually get Rs. 2.25 instead of Rs. 2, if you let your money be compounded by 50% every 6 months every 6 months instead of 100% every year. Quite interesting!

Does that mean that if we do this more frequently, we can trick the bank into giving us infinite money?

We tried increasing the money by 50% (i.e. ½ of original amount), every 6 months (i.e. 2 times every year). What if we increase 1/12 of the original amount 12 times a year? Is that any better? Let us try:

For the first month, we get:

$$Rs.\left(1+\frac{1}{12}\right)$$

For the second month, we get:

$$Rs. \frac{1}{12}\left(1+\frac{1}{12}\right) + Rs.\left(1+\frac{1}{12}\right) = Rs.\left(1+\frac{1}{12}\right)\left(1+\frac{1}{12}\right) = Rs.\left(1+\frac{1}{12}\right)^2$$

For the third month, we get:

$$Rs. \frac{1}{12}\left(1+\frac{1}{12}\right)^2 + \left(1+\frac{1}{12}\right)^2 = \left(1+\frac{1}{12}\right)^2\left(1+\frac{1}{12}\right) = \left(1+\frac{1}{12}\right)^3$$

The calculation gets too tiresome by the end of 12 months, but then, now we see the obvious pattern in the formulas.

For 12th month, we get:

$$Rs.\left(1+\frac{1}{12}\right)^2 = 2.61303\ldots$$

Well, not bad! We got Rs. 2.61303…instead of just Rs. 2.25, if we allow the money to grow 1/12 of the original amount 12 times a year. What if we get let it grow 1/52 of the original amount every week of the year, at the end of the year, we would get:

$$Rs.\left(1+\frac{1}{52}\right)^{52} = 2.6925\ldots$$

If we let is grow 1/365 of the original amount every day of the year, then, at the end of the year, we would get:

$$Rs.\left(1+\frac{1}{365}\right)^{365} = 2.7145\ldots$$

There are about 31536000 seconds in a year, so what if we let the money grow every second? Then, at the end of the year, we would get:

$$Rs.\left(1+\frac{1}{31536000}\right)^{31536000} = 2.7182\ldots$$

Did you notice something really sad happening? We aren't growing much now! We barely grew from Rs. 2.7145…to Rs. 2.7182…even if we compounded our money every single second! It seems that nature has somehow put a limit to how much something can grow! That no matter how hard we try, we will be stuck at some number. This is sad, really sad and fascinating at the same time. Now, one must realize that we could compute powers like 365 or 31536000 here in our example with the help of calculators and advanced computing techniques. But Jacob really had no access to these things. So even though he could never figure out the number, he said it must be between 2 and 3. Bernoulli called it the *force of interest*. A sort of stopping value to the growth.

Somehow, the nature is in favor of banks and says to us, "Thou shall not trick the banks and have infinite money!"

3.2 The Euler's Number: e

We sort of generalize the formula that if something grows by 1/n of its original amount, n times in a given time, then the final answer is of the form:

$$\left(1+\frac{1}{n}\right)^n$$

So, if something grows, every millisecond, every nanosecond, in fact, every instance; what will the above expression be? n will be a very very large number (remember we called it ∞). The goal is simple now, we keep putting in larger and larger numbers into n, and we get a value. We did the same in previous section with a few numbers – that should have given you the idea. Obviously, the numbers has no end i.e. it is irrational. That irrational number is called e, Euler's number! i.e.:

e = 2.71828182845904523536028747135266249775724709369999595749669676277...

But hold on! If Jacob did all the work, why is the number named after Euler? After all, who has he even? **Warning:** If you do not know who Euler is, close the book right now and go read about him. He is the greatest mathematician of our times.

Fig. 3.2: Leonard Euler ("Oil"er not "U"ler)

As a young boy, Euler was influenced heavily by Jacob's brother, Johann Bernoulli. It was Euler who actually calculated the value of e to eight decimal places and recognized it as a mathematical constant. There is no specific reason as to why he denoted the number with the letter 'e' (Personally, I think he named the number after himself!). Some say, he used e for exponential; some say he was already using a for many constants, so he picked up the next vowel to denote this constant. Whatever it may be!

The constant e is kind of like an upper limit to growth. Think of it as the speed of the light being the upper limit to how fast you can go. Kind of. It will get better. With time in this book and later in Vol. 2, we shall encounter this number again and again, specifically in Chapter 8 of this book. The reason I discuss it here is because in so many mathematical texts, e is introduced as just another transcendental, irrational mathematical constant. People don't know where and how it came into being. I admit that it is my bias that I give importance to anything related to Gauss or Euler!

Any field of math: *exists*

Gauss and Euler:

Meme 3.1: Talk about poking your nose

It is strange, uneasy and senseless at the same time. Why would nature do this? Is nature limited by mathematical constants? Perhaps. I would advise you to digest this fact for now. In fact, whenever growth or decay will be discussed, you will see that e pops up automatically.

And there is no doubt that there are a billion things growing and decaying – including the universe itself!

3.3 The Famous π

Of course, you have seen π before. It is usually introduced while teaching about the area and circumference of the circle. They tell its an irrational number that you can approximate as 3.14 or 22/7 but casually ignore where does it come from. One of the earliest approximations about π comes from Indian mathematician Aryabhatta, who coded it in form of verse (Ganitapada, Verse 10) as shown in Fig. 3.3

चतुरधिकं शतमष्टगुणं द्वाषष्टिस्तथा सहस्राणां ।
अयुतद्वय विष्कम्भस्यासन्नो वृत्तपरिणाहः ॥ Sloka 10.

Fig. 3.3: Sanskrit Verse about approx. value of π

The verse in Fig. 3.3 translates to the following: Add 4 to 100, multiply by 8 and add to 62,000. This is approximately the circumference of a circle whose diameter is 20,000." i.e. π ≈ 62832/20000 = 3.1416, which is a fair approximation. Anyway, in order to give a big picture, it is better to avoid the immensely long history of π and rather deal with a very interesting question to understand it as given in illustration 3.1.

Illustration 3.1: How would your handwriting change if you lived in a universe where π = 3, instead of π = 3.1415?

Pretty weird question. Doesn't even make sense. But hold on! It will.

Meme 3.2: Be Some Irrational at Times!

Let us define a ratio k as follows:

$$k(shape) = \frac{perimeter\ of\ the\ shape}{longest\ diagonal\ of\ the\ shape}$$

Before getting to the actual answer, let us calculate k for three regular shapes, i.e. square, octagon and decagon. Let a be the length of each side of these shapes. Then, diagonal of square is given by $\sqrt{2}a$, the diagonal of an octagon is given $\sqrt{4+2\sqrt{2}}a$ and the diagonal of decagon is given by $(1+\sqrt{5})a$. (Don't worry if you don't know these formulas, you can look at them in any geometry textbook).

Now,

$$k(square) = \frac{perimeter\ of\ the\ square}{longest\ diagonal\ of\ the\ square} = \frac{4a}{\sqrt{2}a} = 2.828427...,$$

$$k(octagon) = \frac{perimeter\ of\ the\ octagon}{longest\ diagonal\ of\ the\ octagon} = \frac{8a}{\sqrt{4+2\sqrt{2}}} = 3.0614674...,$$

$$k(decagon) = \frac{perimeter\ of\ the\ decagon}{longest\ diagonal\ of\ the\ decagon} = \frac{10a}{(1+\sqrt{5})a} = 3.0901699...$$

Here is an observation: As we increase the number of sides, the ratio k for that shape starts approaching a limiting value of "3 point something-something" and this ratio k somehow gives us an idea that given a number how many sides would allowed to the polygon. Quite simple. Now, how about k (circle)? The number of sides has to approach infinity (which gives us an idea that this ratio will never be accurate enough and will continue forever). Also, perimeter will be called circumference and

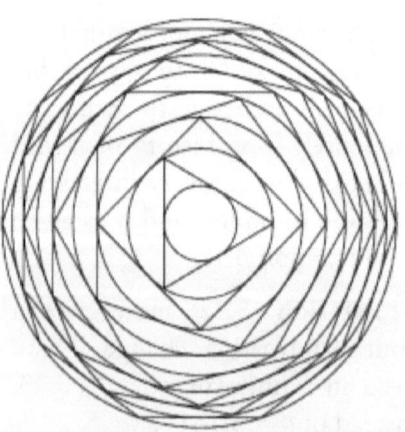

Fig. 3.4: Circle as a polygon of infinite sides

longest diagonal will be called diameter. Simple. Also, k(circle) is what we famously term it as π. Something like shown in Fig. 3.4.

$$k(circle) = \frac{perimeter\ of\ the\ polygon\ of\ infinite\ sides}{longest\ diagonal\ of\ the\ polygon\ of\ infinite\ sides}$$

Or, in famous terms,

$$\pi = \frac{Circumference\ of\ the\ circle}{Diameter\ of\ the\ circle} = 3.1415...$$

Thus, it gives us an idea that this ratio k refers to the curvature of the polygon we are referring to. So, when you are approximating $\pi = 3.1415...$ to just 3 you are also approximating circle to a shape whose k is 3. And what is that shape? It turns out that the shape is a hexagon as we can see below:

$$k(hexagon) = \frac{perimeter\ of\ the\ hexagon}{longest\ diagonal\ of\ the\ hexagon} = \frac{6a}{2a} = 3$$

When we write, we use a lot of curvature (as you can see in everything written in this paper). All these curvatures happen to be part of some or the other circle. But physically if π was actually 3, then the curvatures that we would be allowed would be part of a hexagon and never a circle. Thus, no cursive handwriting would be possible. All we would be allowed to write would be sections/parts of hexagon. Can you draw the alphabets now? (Remember they have to be sections of a hexagon).

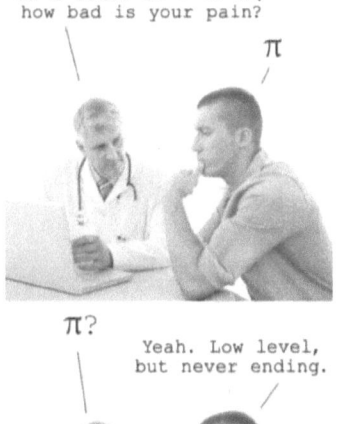

Meme 3.3: Ain

3.4 Imaginary Number *i*

To any civilized mathematics student – this will simply sound outrageous! What do you even mean by imaginary numbers?

As if transcendental, irrational and several "real" numbers were already not enough – now we have to deal with imaginary numbers too.

An imaginary number is defined as $i = \sqrt{-1}$ (square root of negative one!). The symbol i for imaginary numbers was again chosen by Euler. I would have loved to expand on the history and physical meaning of i, but fortunately we won't be using it anytime soon in our course of precalculus and calculus. But I encourage you to learn more about them.

Real numbers and imaginary numbers combined together are called Complex numbers. Luckily, we shall only restrict ourselves to Real numbers and real number line. However, there exists a graph of complex numbers as well in a plane called Argand Plane. The name imaginary is misleading. It does not mean at all that they do not have any physical significance. Although hard to imagine, complex numbers form much of the modern world that we have today.

Meme 3.4: Matrix reloaded

3.5 The Most Beautiful Equation in Math

Euler combined all these five mathematical constants in one equation – often called as the most beautiful equation in all of mathematics, as shown:

$$e^{i\pi} - 1 = 0$$

No one has ever understood this equation, but yet we know that it is true. Just imagine an irrational number e raised to an imaginary power and that raised to an irrational power of actually gives you a real number! How crazy is that?

Later in our course in Calculus in Vol. 2, we shall attempt to prove Euler's equation.

3.6 Closure

This chapter was purely intended for your personal reading. It is necessary to get comfortable with the numbers you'll be using often and to truly understand where they come from. Hence, there is no intention of making exercise questions to this chapter. However, feel free to know more about them.

4 Decoding the Mind of Nature: An Introduction to Functions

"Beauty is the first test: there is no permanent place in the world for ugly mathematics."

– G. H. Hardy

We pick this section up from sort of the rough ideas we had from Chapter 2. Kind of building upon those stuff (In case you need to review).

Look around. At first, nature seems pretty random and chaotic. In fact, to ancient people, everything around them would seem to be stuff made up of magic. The seasonal changes, days and nights, stars and their motion, the very phenomenon of life and death, the sheer vastness of land and the ocean, earthquakes, volcanic eruptions, floods, diseases and epidemics and just what not. There would have been no seemingly possible way to connect these dots. To make sense of this nonsensical chaos.

Then slowly, things came together. We started recognizing small but crucial patterns in nature. For example, the understanding that seasonal changes *"repeat"* themselves after a certain amount of time made it possible for us to be prepared for them. We started counting days and nights *"depending upon"* the number of times this yellow ball of fire has gone in and come out of the sky. We started to make sense of direction *"depending upon"* where stars are in the night sky. We discovered agriculture and figured that the good crops will be produced *"depending upon"* the amount of sunshine, rainfall, quality of soil etc. We started civilization and moved around in carts that moved *"depending upon"* how circular/smooth their wheels have been shaped.

Cut to modern day – all modern devices *"depend"* on electricity. Our travel *"depends"* on the mode of transportation that we choose to take. The speed of a computer *"depends"* on its RAM, storage unit etc. Almost everything in nature *"depends"* on time. In fact, all of life *"depends"* on the choices that we make (like the one to pick this book, I appreciate it!).

It seems to mankind that the key to understanding nature is to understand *"dependence"*. To understand any phenomenon or magic happening around us, we need to figure out what that thing depends on. What controls it? How many things control it? And how do they change? In fact, if we understand the *"dependence and how they change"*, we have pretty much cracked the code to speaking to the nature!

With this philosophy in mind, we start the first baby steps to understanding Calculus – this complex mechanism of "dependence" in the language of mathematics or what we will call in the words of Gottfried Leibniz as – ***"functions"***.

4.1 Leibnitz and the Algebra of Thought

To get started on this chapter, let's get into the mind of this crazy person who coined the word "Functions" – Gottfried Wilhelm Leibniz (who by the way coined many other words that we will constantly use now, such as *constant, variable* and *parameter*). Other than the famous fight between Newton and Leibnitz about who invented Calculus, this dude had pioneered Mathematics, physics, medicine, biology, psychology, engineering, history, politics, diplomacy, logic, poetry, ethics and even veterinary medicine of all the things! (I bet he would be an ideal kid to Indian parents in 21st century).

Anyway, like many mathematicians before him, he was obsessed with symbolism. This idea is perhaps not old. In fact, about 1400 years before him, Diophantus (an Egyptian mathematician, also known as the *father of algebra*) made first attempts to translate verbal mathematics into symbols.

Exemplum 4.1: Let's use symbolic idea used by Diophantus to solve a riddle about his age by Metrodorus:

'Here lies Diophantus,' the wonder behold.

Through art algebraic, the stone tells how old:

'God gave him his boyhood one-sixth of his life,

One twelfth more as youth while whiskers grew rife;

And then yet one-seventh ere marriage begun;

In five years there came a bouncing new son.

Alas, the dear child of master and sage

After attaining half the measure of his father's life chill fate took him. After consoling his fate by the science of numbers for four years, he ended his life.'

Fig. 4.1: Diophantus

Solution: This is a pretty long, boring attempt at poetry. But let's see if it translates to symbols. Let us denote his total age/life by the '*variable*' called 'x' and now we translate line by line.

    a. God gave him his boyhood one-sixth of his life, which means it is $\frac{x}{6}$.

    b. One twelfth more as youth while whiskers grew rife, which means it is $\frac{x}{12}$.

    c. And then yet one-seventh ere marriage begun, which means it is $\frac{x}{7}$.

d. In five years there came a bouncing new son, literally can be represented by 5.

e. After attaining half the measure of his father's life chill fate took him, which means it is $\frac{x}{2}$.

f. After consoling his fate by the science of numbers for four years, he ended his life, is also literally the number 4.

By combining all of it, this poetry, line by just means that his total life (x) was the addition of the following parts:

$$\frac{x}{6}+\frac{x}{12}+\frac{x}{7}+5+\frac{x}{2}+4$$

That is, literally the entire riddle/problem is $\frac{x}{6}+\frac{x}{12}+\frac{x}{7}+5+\frac{x}{2}+4 = x$. Or,

$$\Rightarrow \frac{x}{6}+\frac{x}{12}+\frac{x}{7}+5+\frac{x}{2}+4 = x,$$

$$\Rightarrow \frac{14x+7x+12s+420+42x+336}{84} = x,$$

$$\Rightarrow \frac{75x+756}{84} = x,$$

$$\Rightarrow 9x = 756, \text{ or } x = 84$$

Hence, Diophantus died at the age of 84 years according to the riddle.

This is the magic of mathematical symbolism. It can simply boil down lines and lines of ideas in probably inches of space. On a lighter note, I think this is what gives mathematics its "complex, irritating and impossible to understand" impression. But really, each variable is simply a lazy way of writing long, long set of words. Mathematicians are okay with being lazy and looked upon as difficult beings.

Leibniz thought that most *"dependence"* in nature and in the world around us can be symbolically written by *"variables"* such as x, y, z etc. And seen in Exemplum 1, when things get complicated and really long, symbols gives us a way to see them mathematically and in a short and concise manner. So, when we say, the perimeter of a square is four times its side length; we can simply write $P = 4a$, where P represents perimeter of square and represents its side length. Notice that is simply a lazy way to show "dependence" of perimeter of square on its side length with the help of algebraic and mathematical symbols.

I guess it's quite obvious to note that the value of P will depend on the value of a that we choose. Hence, we say P is the **dependent variable** (*where we do not have choice over its value*) and a is the **independent variable** (*where we are free to choose a value for it and that in turn affects the value of something else*).

A little change in the language: Instead of saying P *depends* on a, we will now start saying that P is ***function*** of a. Or more precisely, P is a **single variable** *function* of a (as it depends on only one quantity). We started using the word "function" without actually declaring what it really is, but don't worry – right now I just wanted you to start making it a part of your language, we shall see the actual meaning and symbols in the next section. Few more examples:

a. Volume of a cube (V) depends on its side length (x) with dependence being that volume is cube of the side length. In symbols, $V = x^3$ and is a *single variable* function.

b. Volume of a cylinder (V) depends on its radius (r) and its height (h) with dependence being that volume is π times the square of radius times the height. In symbols, $V = \pi r^3$ and V is a function of *two variables*.

c. Volume of a cuboid (V) depends on its length (l), breadth (b) and height (h) with dependence being that volume is length times breadth times height. In symbols, $V = lbh$ and V is a function of *three variables*.

(By the way, remember Descartes? The guy from Cartesian Plane. It was he who decided to use the symbols x, y, z for things that are unknown and the symbols a, b, c to denote the known numbers or *constants*).

Query 4.1: *Let's return to the title of this section – "Leibnitz and the algebra of thought". Leibnitz believed that he could possibly express everything in this world including human thought by mathematical symbols* *and expressions. Do you think that given the complexity of emotions and human thoughts, it is possible to express them in the form of mathematical formulas or functions of some set of variables?*

I hope you get rough hold of it. Returning to our introduction, try reading it again by replacing the word "dependence" with "function" (minor changes wherever required, not literal). Well, enough of chit-chat! Let's get started with the formal definition of function.

4.2 Lobachevsky-Dirichlet Definition of a 'Function'

As with any term/concept, there are a lot of mathematicians over the ages who have come up with different definitions of what a function really is. Starting from Issac Newton(1713), Euler(1748), others to all the way till Weierstrass(1861) and Bourbaki(1939). The older ones are perhaps easy to understand but not really correct. The modern definitions, however accurate they are, become too abstract and complicated to understand in the first go. Hence, I have decided to take the mid-way path and combine the definitions given by Lobachevsky(1834) and Dirichlet(1829), which is what is usually given to first-time learners of Calculus.

"If there is a unique quantity, say represented by variable y that depends and varies gradually with change in all values of another quantity represented by variable x over a given interval, then y is said to be a function of x and is represented by $y = f(x)$. The relation between the quantities can of numerical, analytical or graphical nature or may exist and remain unknown."

Don't worry if it sounds difficult – it is just fancy, mathematical language. Now, instead of explaining this word by word, I would rather show you an illustration of this definition (and this I do, purely out of my own bias and excitement) with the help of a game-changing idea by the 'father of modern physics' – *Galileo Galilei*. Galileo was one of the first persons to mathematically interpret nature. To somehow try to find if nature was following some kind of 'mathematical rules'. Having spent half his life proving that Earth actually moves, he focused on

Fig. 4.2: The Great Galileo

something quite basic towards the end of his life – *how things fall?* Here goes the illustration:

Illustration 4.1: Galileo and falling objects:

We all know that the distance travelled by an object 'depends' on the time it has spent travelling. But how exactly? In daily life, the motion involving falling objects happen so quickly that we are unable to consciously measure the time it takes to travel any unit of distance. Galileo thought that one way to get rid of this problem was to slow down the motion and then check the distance that a particular object travels per unit of time. This is where the inclined plane comes into the picture as shown in Fig. 4.3.

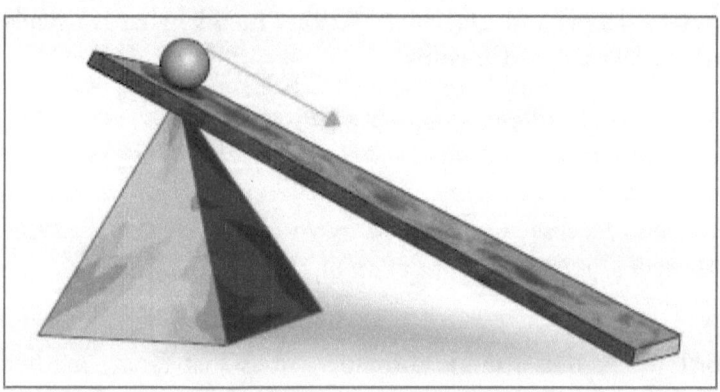

Fig. 4.3: Inclined slope to measure the speed of the ball

This kind of arrangement made it possible for him to slow down the motion and observe how much distance a ball would travel rolling/falling down. Now, for a change – we will not repeat the exact experiment here (i.e. we are not using the exact measurements used by Galileo). *It is the idea that is important.* With our modern-day equipment, it is not impossible to imagine a device that can measure distance and time super-accurately. Using a computer simulation, we have the following data about an object freely-falling on the surface of the earth as shown in Table 4.1.

Table 4.1: Distance vs. Time of a falling object

Time (seconds)	Distance (meters)
1	4.9
2	19.6
3	44.1
4	78.4
5	122.5
6	176.4
7	240.1
8	313.6
9	396.9
10	490

So, we have measured distance traveled by a falling object for 10 seconds over a gap of 1 second by computer simulation. Let us call the time as the variable x (although t would have been better) and distance traveled as y. Observe, carefully. The distance y is a "function" of time (meaning y gradually changes as x changes), written in short hand as $y = f(x)$. For each value of x (i.e. at every point of time), there is "unique" value of distance y traveled (would it not be completely nonsensical if object traveled two different distances at one point in time?) Hopefully, now some part of Lobachevsky-Dirichlet definition has started to make sense.

Notice that at this point, although we know that somehow the distance is dependent on time; we don't really know "how" they are related to each other. We just have this dependence in form of *raw numbers*. This kind of representation of functions is called *"numerical representation"*.

Let us try to find some patterns in these numbers. For starters, and it may not be very obvious that all the values of (distance traveled) have 4.9 as a factor. We now, re-construct our table, as shown in Table 4.2.

Table 4.2: Seeing 4.9 as a factor

Time (seconds)	Distance (meters)	Distance (meters)
1	4.9*1	= 4.9
2	4.9*4	= 19.6
3	4.9*9	= 44.1
4	4.9*16	= 78.4
5	4.9*25	= 122.5
6	4.9*36	= 176.4
7	4.9*49	= 240.1
8	4.9*64	= 313.6
9	4.9*81	= 396.9
10	4.9*100	= 490

Interesting! The next should be now easy to spot! After removing 4.9 as a factor, the left-over numbers are simply squares of the time that they *correspond* to. Numerically written as shown in Table 4.3.

Table 4.3: Finding all patterns

Time (seconds)	Distance (meters)
1	$4.9*(1^2)$
2	$4.9*(2^2)$
3	$4.9*(3^2)$
4	$4.9*(4^2)$
5	$4.9*(5^2)$
6	$4.9*(6^2)$
7	$4.9*(7^2)$
8	$4.9*(8^2)$
9	$4.9*(9^2)$
10	$4.9*(10^2)$

And there we go! It now seems that we don't need to draw this table at all! We have established a pattern, which we write as follows:

$$y = f(x) = 4.9x^2$$

So the game is very simple, you plug the value of x (time), this relation asks us to square that time and multiply it by the *constant* 4.9 and we can get our distance. In fact, this lets see beyond what is given. We can even plug time x as 11 seconds, or 12 or 1000 seconds and yet expect an answer out of it! This kind of representation of function where we write it in the form of a formula/equation is called the *"analytical representation"*.

I would like to ask you to pause for a moment and feel how Galileo would have felt when he would have discovered that the dependence of distance traveled by the falling objects is proportional to the square of their time period! To cover this immensity of falling in a simple equation is no less than an emotion of deep salvation. To be finally be able to understand what nature is saying. To pave the path for the generations of thinkers and learners to come by.

Recalling what we learned in Chapter 2, Descartes formed a bridge between Algebra and Geometry and hence the dependence $y = f(x) = 4.9x^2$ can even be plotted in the form of a graph for us to see a 'picture' on Cartesian Plane as shown in Fig. 4.4:

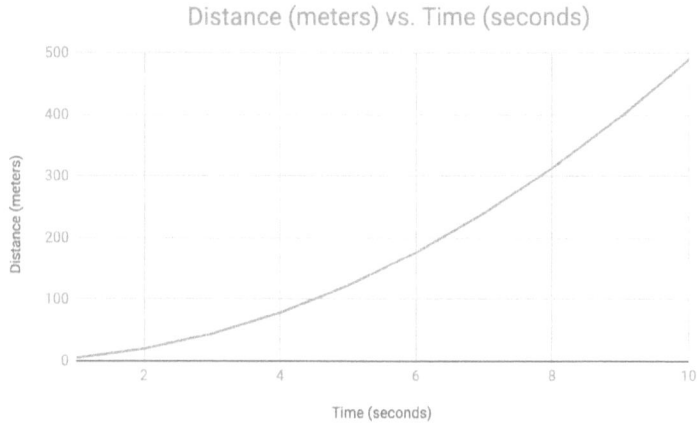

Fig. 4.4: Graph of dependence of $y = f(x) = 4.9 \, x^2$

This visual way of representing our dependence/function is what we call as the *"graphical representation"*. Also to mention as bonus that x is our independent variable here and y is the dependent variable.

Look back and read the Lobachevsky-Dirichlet definition of a function again. I hope the idea in each word of the definition is more or less explained by this illustration (although not completely). It's been a remarkable progress starting with symbols to represent things to actually making dependence a mathematical trick to unlock mind of nature.

Query 4.2: Some guy measured the speed of an object as it falls down in a vertical tunnel in vacuum. He collects the following data:

Distance traveled in Tunnel (m)	10	40	90	160	250	360	490	640
Velocity (m/s)	14	28	42	56	70	84	98	112

For the given numerical representation of function, find the analytical and graphical representation. Assign variables to the quantities as per your wish.

Meme 4.1: Can a function be an input for another function?

4.3 Functions as Mapping from Sets to Sets

The sole purpose of this section is to make you familiar with the way functions are written and understood in modern mathematics. Just kind of staying up to date. Mind you, we are not adding any new magical concepts here. We are merely adding a few more fancy words, ideas to our vocabulary and changing the picture we are used to painting.

Let us take the numbers we had in Illustration 1 for understanding what is going on. One can try drawing a diagram as shown in Fig 4.5:

This should not be too hard to interpret. We got a bunch of values of time (x) ranging from 1 second to 10 seconds and a bunch of value of distances (y). We figured out the dependence as $y = f(x) = 4.9x^2$ and each value of time (x) was "*mapped*" to some value of distance (y).

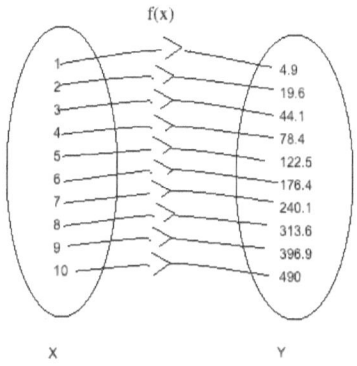

Fig. 4.5: Arrow diagram of dependence of $y = f(x) = 4.9\,x^2$

We can write the same idea in mathematical symbols as:

$$f : X \to Y$$

Which literally means that we can imagine a bag X containing values of time, a bag Y containing value of distances and the dependence/rule/function ($f(x)$) that told us which value of time(x) should be mapped to what exact value of distance (y).

Now, needless to say, we will not be dealing with time as (x) and distance as (y) in all cases. In fact, they can be any two quantities that have a dependence. The abstract idea is important here. The symbol $y = f(x)$, was introduced by Euler and it helps to see that the independent variable x affects the value of y by going inside (or becoming *argument* of) the function f. Something like shown in Fig. 4.6.

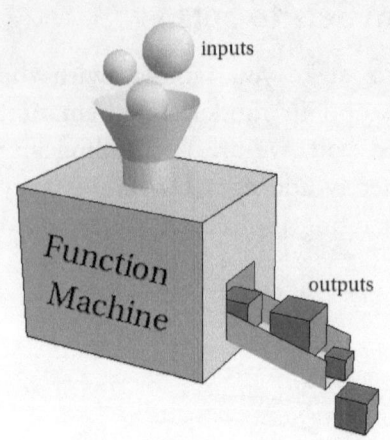

Fig. 4.6: Function as a machine

Let us focus on the bag of values where we get input/independent variable x from. It is called the *Set X* in mathematics. Does it have restrictions? Well, at least in the example in Illustration 1, there definitely is. We cannot have time physically as a negative quantity, it needs to be greater than or equal to zero (symbolically $x \geq 0$). In a nutshell, only values $x \in [0, \infty)$ are allowed (I assume you are okay with this notation from sets, if not please refer to any material containing how sets are represented). Then, the set is what we call the ***domain*** of the function (Simply put, the domain tells you what values you can use in a dependence or function).

The bag of values where get answers from putting the value in function, denoted by *Set Y* is what we call the ***range*** of the function. From Illustration 1, we can say that since there are no negative values of time (or *Set X*) and the function is $y = f(x) = 4.9\ x^2$, we do not expect any negative answers as range also (Simply put, domain tells you what values you can expect out of a dependence or function). Simple ideas.

This idea of domain and range will probably come in every chapter of this book now. So even if it sounds a bit unclear or hazy, don't worry! You'll see it enough number of times till a point where it is crystal clear.

Finally, two very important restrictions to remember that make a function a valid one:

a. *Every element in X must be mapped to a unique element in Y.*

b. *Every element in X must have a mapping.*

We will see both these ideas not in terms of words but rather in pictures as shown in Fig. 4.7.

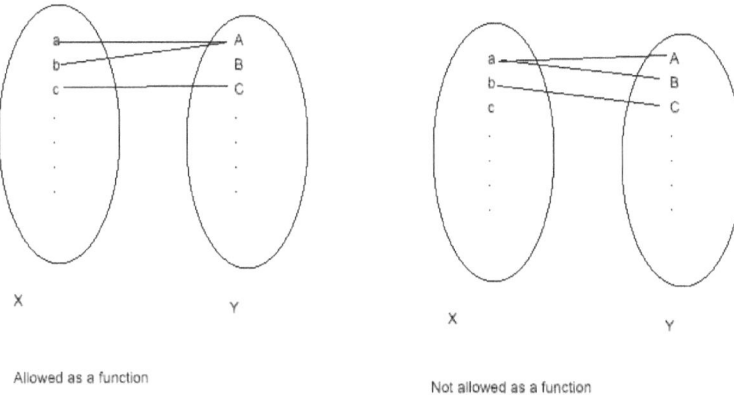

Fig. 4.7: What is and what is not a function

To conclude this section in the words of Nicolas Bourbaki (1939):

Let X and Y be two sets, which may or may not be distinct. A relation between a variable element x of X and a variable element y of Y is called a functional relation in y if, for all x an element of X, there exists a unique y an element of Y which is in the given relation with x.

4.4 Implicit and Explicit Functions

Let us say we have a function $y = f(x)$ defined by $y = x^3 + 4x + 5$ here it so happens that all the terms containing the independent variable and all the terms containing the dependent variable are on two different sides (isolated). These are called *"explicit" functions*.

On the other hand, if the independent variable and the dependent variable are not separated but are mixed up in an equation, let us for example, $2xy + y^2x + \sqrt{x} = 5$, these kind of functions are called *"implicit" functions*. That's it! Bunch of words to add to your mathematical vocabulary.

4.5 The Tree of Functions

After a long load of technical words – let us wonder about something rather very basic. We started this whole discussion by the word "dependence". Exactly how many kinds of these dependencies are there?

Are they simple? Do all dependencies require only a handful of variables to work? Can we really find all these dependencies at all times in all given situations in all experiments?

The answer to all of the questions above, unfortunately is – **NO**. See, make it as romantic as you want, but nature is complex. Think of the weather – it depends on crazy amount of things and no matter how accurate the predictions are – we still leave room for doubt. Think about motions of planets, moods of people, spreading of disease, your health, evolution of organisms, evolution of the planet itself and so many other things – they are all functions of so many things changing with each other in an enormously complex manner.

The paintings of nature are quite complex for us to imagine – its colors uncountable. By the way, colors! They remind me of something. Given how much our society discourages us from going into arts – I do not expect many of you guys to painters but I hope you do know this basic fact; all colours that we use in an artwork are just derived out of three colors, called the Primary Colors – Red, Yellow and Blue. You combine them (Red + Yellow = Orange, Yellow + Blue = Green, Red + Blue = Purple) and you get Secondary Colors. Then you mix secondary colors with each other and get tertiary colors and so on and on and on. That means when you are looking at a complex piece of artwork – any new color that you see must hold its origin to some kind of crazy combinations of red, yellow and blue again and again with differing shades. Why do I mention this here?

Lucky for us, we have figured out something what I call as a tree of functions, which loosely behaves like the colors I mentioned above. Any complex function that we figure out in nature have to combination of some "basic functions". These basic functions I have shown in Fig. 4.8. Note that, certainly these are not the only functions available – but rather these are the basic ones. The functions (and their subtypes) that are shown on tree of functions are as follows:

1. Algebraic Functions
    a. Polynomial Functions
    b. Rational Functions
    c. Irrational Functions
    d. Special Functions

2. Transcendental Functions

    a. Trigonometric Functions

    b. Inverse Trigonometric Functions

    c. Exponential Functions

    d. Logarithmic Functions

The entire part of the rest of the book will be dedicated to understanding these basic functions! And I advise you to have really really have a strong grasp on each of them. Try to make sense and feel their graphs to an extent where you should be able to see each one of them in front of your eyes.

Now, for the sake of definitions, what are algebraic and transcendental functions? An algebraic function, as the name suggests – involves only operations defined in algebra, like addition, subtraction, multiplication, division, fractional exponents etc. A transcendental function is any function that is not algebraic (for example, logarithmic, trigonometric etc.). That's all.

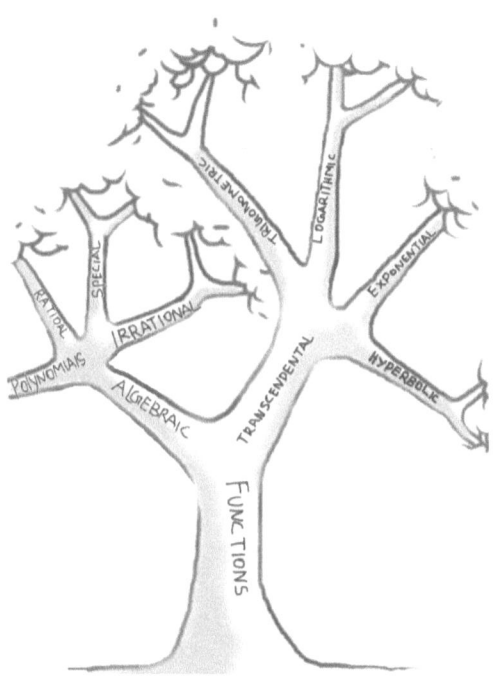

Fig. 4.8: The holy tree of functions

4.6 Closure

We shall be discussing each of the branch of tree of functions one by one. Polynomial functions are discussed in Chapter 5. Rational and Irrational Functions are discussed in Chapter 6. Trigonometric Functions are discussed in Chapter 7. Exponential and logarithmic functions are discussed in Chapter 8 and finally some special functions and inverse trigonometric functions are discussed in Chapter 9. So, in a way, the next 5 chapters that follow are basically about mastering these branches of the tree of the functions. Chapter 9 also is about how to combine these basic primary colors to make more complex secondary and tertiary colors and of course, Chapter 10 is all about making some paintings!

Exercises

1. Given a function $f(x) = ax^2 + \dfrac{a}{x^2} + bx + \dfrac{b}{x}$, where a and b are constants. Prove that $f(x) = f(\dfrac{1}{x})$.

2. The surface area S of a sphere of radius r is given by $S = 4\pi r^2$ and the volume V is given by Express $V = \dfrac{4}{3}\pi r^3$ as a function of S, $V = f(S)$.

3. Can a function act like an input into another functions? If yes, give examples. If not, why?

4. The mirror formula is given by:

$$\dfrac{1}{f} = \dfrac{1}{v} + \dfrac{1}{u}$$

Where f is the focal length, v is the image distance and u is the object distance.

Suppose we have a mirror of focal length f = 5 c.m. Express image distance as a function of the object distance and draw the graph/numeric representation.

5. Can a function act like an input to itself? If yes, give examples. If not, why?

6. Can the points (1, 2), (2, 6), (−111, 65), (1, 17), (41, 66) and (66, 96) be part of any function?

7. Express $x^3 + y^3 = a^3$ in explicit form.

8. A variable x is directly proportional to a, a is inversely proportional to b, b is proportional to the third power of c, c is inversely proportional to one-fifth power of d, d is directly proportional to y. Form a function that shows the dependence on y on x, $y = f(x)$. Draw the graph of the function by taking a few points. (You may take the constant of proportionality as 1 everywhere for the sake of simplicity).

9. Let the height of a right-angled triangle be x and its base be y. For a fixed length of hypotenuse of 5 units, use Pythagoras theorem to find an implicit function. Show all the possible heights and bases. Graph the obtained implicit function. Does the shape obtained on the graph have some relation to triangles?

10. Let there be two functions f and g. $f(x) = 3x+5$ and $g(x) = x^2 + 6$. Values from 1 to 10 (natural numbers) are used as input in *f(x)* and the obtained output obtained is used as an input in *g(x)*. Draw an arrow diagram representing the whole process.

11. Does the equation $y^2 - x = 47$ represent a function, either $y = f(x)$ or $x = f(y)$?

5 To Visually See the Algebra: Polynomial Functions

"For the things of this world cannot be made known without a knowledge of mathematics."

– Roger Bacon

To begin with, it must simply be stated that algebraic functions are the *easiest* functions to deal with (a good and a bad news). The Greeks, Egyptians and the Indians usually liked to deal with algebraic functions in form of geometry. Most of them liked to imagine x^2 as the area of a square whose side x was x^3 and as the volume of a cube whose side was x. But I guess, the story has to end there – because what would you imagine $x^4, x^5,\ldots$ as? Something related to a 4,5,…dimensional shape? Can we even imagine that? And this is where you will realize that *the power of algebra begins where geometry ends* (in order to understand this, look at the Illustration 5.1).

Illustration 5.1: Let us consider a rectangle with length and breadth as x_1 and x_2 respectively. Suppose, we wish to find the length of the diagonal of the rectangle. Then, by the formulas of elementary geometry:

$$\text{Diagonal of a rectangle} = \sqrt{x_1^2 + x_2^2}$$

Let's move to a three-dimensional shape now, probably, a cuboid of length, breadth and height as x_1, x_2 and x_3 respectively. What now is the diagonal of the cuboid? Pretty simple.

$$\text{Diagonal of a cuboid} = \sqrt{x_1^2 + x_2^2 + x_3^2}$$

You should be noticing a trend here, aren't you? The square of the extra dimension simply goes inside the square root in the formula of the diagonal. Now, think of something unimaginable. *What would be the diagonal of a four-dimensional figure having dimensions as x_1, x_2, x_3 and x_4?*

With the three-dimensional figure like cuboid, you could at least have an imagine! Think what the diagonal looks like. But here there seems to be no image. That is the difference between a mathematician and common people. A mathematician won't care about the image – he will simply study the pattern and write:

$$\text{Diagonal of a 4-dimensional shape} = \sqrt{x_1^2 + x_2^2 + x_3^2 + x_4^2}$$

While the world will keep trying to make sense and discover what it means centuries later, a mathematician is already there – well before anyone else and can even extend the formula to n dimensions.

To get started with the first branch of algebraic functions, i.e. polynomial functions, I will have to take you to a special part of the dark ages of Europe (476–1492 A.D.) known as the Renaissance period (1300–1600 A.D., meaning *rebirth*) and its connection with polynomials (as if algebra wasn't dark enough!).

5.1 "Dark Ages" and Polynomial Functions

In fancy symbolic language of mathematics, a polynomial function would be defined as follows:

$f(x) = a_n x^n + a_{n-1} x^{n-1} + a_{n-2} x^{n-2} + \ldots + a_2 x^2 + a_1 x^1 + a_0$, where $a_n, a_{n-1}, \ldots, a_1, a_0$ are real numbers and is a positive integer… **Def (i)**

Polynom-nom-nom-nomial

$+ x^2 + x^3 + x^4 + x^5 + \cdots$

Meme 5.1: Poly"nom"ials

All the symbols that you see here must be quite familiar to you. They appear in almost every boring textbook of algebra that you come across. What may be unknown to you is their enlightening origin.

The dark age of Europe is defined as a period of ignorance. A place where free-thinking was restricted and where questioning was forbidden. Where to that part of world – flat earth was the ultimate reality and no one was supposed to *reason*. People like Bruno were burned by Church for saying that the universe was infinite and that we are not the center of it. Political instability and diseases were widespread. But then, the human spirit is unshakable and relentless – isn't it?

That very relentlessness pushed them to Renaissance Era – where people started to believe in their own power of reasoning and logic which gave us the *modern symbolism of mathematics*. Look at every symbol in Def (i): '+' the sign introduced by *Nicholas Oresme* (14th century mathematician who was one of the first to challenge authority of Aristotle), the '–' sign introduced by Widmann (15th century mathematician who was one of the first to give a university lecture on algebra), the '=' sign coined by Robert Recorde (born in 1512 who died in prison after being accused of being "political enemy") to the very idea of using letters like *a, b, c,…,x, y, z* in algebra instead of words by Vieta (who was appointed by French King to break the enemy's secret code in times of war); all of them came from this period. The entire look of modern mathematics came in this time of awakening and that makes me wonder that these are *not just symbols of mathematics alone.*

Coming back to Def (i), look at the first term– $a_n x^n$, n indicates the *highest power* of our variable x and for a function to be a polynomial function, n must always be a positive integer. The number a_n is simply a *constant* multiplied to x^n and is called its *coefficient* (another fancy term), which basically means that for a given function it'll be fixed. It is assumed in algebra that once a power of x^n exists, the lower powers and their corresponding coefficients will also exist. And if they don't, we can always imagine their coefficients to be 0. Let me simplify this for you.

For example, if $f(x) = 2x^4 + 5x$, this might look like a polynomial of degree 4 for but has no terms involving powers 3, 2, 1, and 0. But we can always re-write it down as:

$f(x) = 2x^4 + 0x^3 + 0x^2 + 5x + 0x^0$, where $a_4 = 2$, $a_3 = a_2 = a_0 = 0$ and $a_1 = 5$. As simple as that! It's a gentleman's habit to imagine polynomial always like this in your head even if it is not written in that form. You see a polynomial with highest power of n, you automatically start imagining the terms containing lower powers like $n-1, n-2, ..., 3, 2, 1, 0$. By the way, the term with the highest power, in our case $2x^4$, is called the *leading term* of the polynomial and its coefficient is what we call as the *leading coefficient*.

Our old, sick fellow Descartes decided that the letters from the beginning of the alphabet x, y, z will be used to write *constants* in a function and the letters from the end of the alphabet will be used to write *variables*. Now, back to the power business, the value of highest power n can be either zero, an even number or an odd number (that's the simplest division we can do, given that we already mentioned that we are not allowing negative or fractional powers). *Let us for now only take the term of highest power and not the lower terms.* Therefore,

a. When $n = 0$ the polynomial function simply is $f(x) = a_0$, which will as a *constant polynomial function*.

b. When $n = 1$ the polynomial function simply is $f(x) = a_1 x$, which will call as a *linear polynomial function*.

c. When $n = $ *some even number (or you may write as 2k, where k is some positive integer)*, the function is simply $f(x) = x^n = x^{2k}$, which we call as a *family of even-power polynomial functions*.

d. When $n = $ *some odd number (or you may write as 2k + 1, where k is some positive integer)*, the function is simply $f(x) = x^n = x^{2k+1}$, which we call as a *family of odd-power polynomial functions*.

Looks quite a bit of technical jibber-jabber here. But it's how algebra starts off as – with boring set of definitions, formulas but ending up to defining almost all changes in nature. Just hold on for a bit. We now look at each of the three categories that I wrote above in detail.

5.2 The Constant Polynomial Function

In the symbolic language, a constant function is simply written as:

$$y = f(x) = a_0 \text{ (analytical representation)}$$

Which basically means the independent variable x, doesn't really affects the dependent variable y; for you can choose any value of x, and y is like eh! I don't really care – I got this a_0 with me and I am good!

Let's look at the mapping diagram of this function as shown in Fig 5.1.

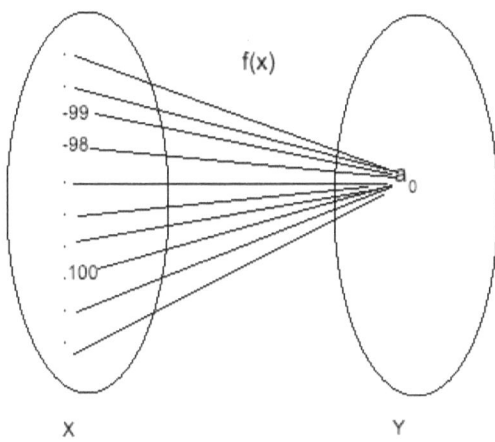

Fig. 5.1: Mapping diagram of constant function

It is pretty obvious that we can use any value of x, that we feel like and hence the Domain of the function is *all real numbers*, **R**. And all those values will be mapped to only one constant a_0 that we chose, hence the Range of the function is just $\{a_0\}$.

A numerical representation of this function would require us to pick a value for a_0. Let's say $a_0 = 73$. (Why did I randomly pick up 73? Because one of my favorite characters Sheldon Cooper from Big Bang Theory says, "The best number is 73. Why? 73 is the 21st prime number. Its mirror, 37, is the 12th and its mirror, 21, is the product of multiplying 7 and 3... and in binary 73 is a palindrome, 1001001, which backwards is 1001001."). The tabular form is shown in Table 5.1.

Table 5.1: Set of values for $y = a_0 = 73$

x	−100	−40	−20	−0.5	0	0.01	1	30	50	100	124	636	635
$y = f(x)$	73	73	73	73	73	73	73	73	73	73	73	73	73

Finally, the graphical representation looks like the one shown in Fig. 5.2

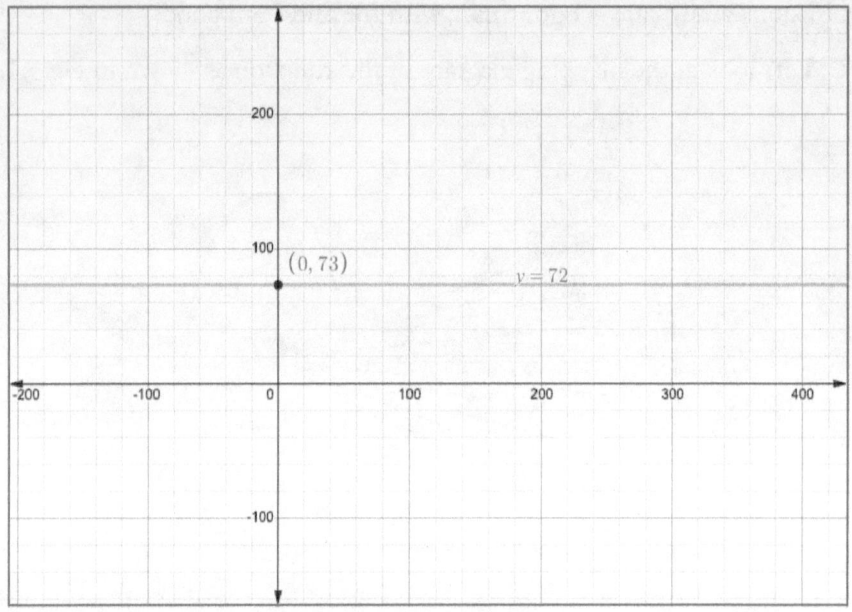

Fig 5.2: Graph of y = 73

Which is basically a line parallel to the x-axis.

Turn up to nature, what things look like they are following this kind of behavior? How about the total energy, E of this universe with time, t ($E = f(t)$)? May be something local, like house-to-house salesman who sell all their products at the same fixed price? (Certainly not the behavior, B of people with time, t).

5.3 Linear Polynomial Function

As we mentioned before, we can write the linear polynomial function as:

$$f(x) = a_1 x \text{ (analytic representation)}$$

What value we must allot to a_1 to get started? Well let's pick up a positive number, may be $a_1 = 7$ (Why 7 randomly again? This time not because it is related to any of my favorite characters, but 7 happens to be the first "Happy number"! If you do not know what happy numbers are, google right away! And yes, there are unhappy numbers too!)

So, we get: $y = f(x) = 7x$, which in numerical representation looks like the one represented in Table 5.2 (by picking random values):

Table 5.2: Set of values for $y = 7x$

x	−2	−1	0	1	2	3
$y = f(x) = 7x$	−14	−7	0	7	14	21

And finally, putting these numbers on the cartesian plane, we will get the graphical representation of the same function as shown in Fig 5.3 (a).

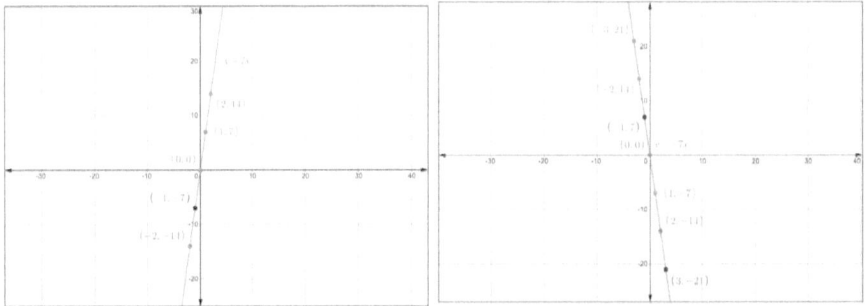

Fig. 5.3: a) Graph of $y = 7x$ b) Graph of $y = -7x$

A nice, neat straight-line graph. Notice one thing, the line makes an acute angle with the positive x-axis, which is another fancy way of saying that as we increase the *x-values, the y*-values will increase too – but in a nice, ordered fashion with equal gaps between them.

Now, how about we put $a_1 = -7$, then our equation comes $y = f(x) = 7x$. We do the same drama again of putting the value and getting the graph as shown in Fig. 5.3(b).

And I hope you get the difference now? The angle has increased to become an obtuse one. Plus, notice one thing: the values of y now actually decrease as x increase but still a nice, neat graph with equal gaps in between the values.

For your own satisfaction, here are graphs (in Fig. 5.4) of $y = f(x) = x$, $y = f(x) = 2x$, $y = f(x) = -x$ and $y = f(x) = -2x$ all in one frame just to make it more clear what a_1 does!

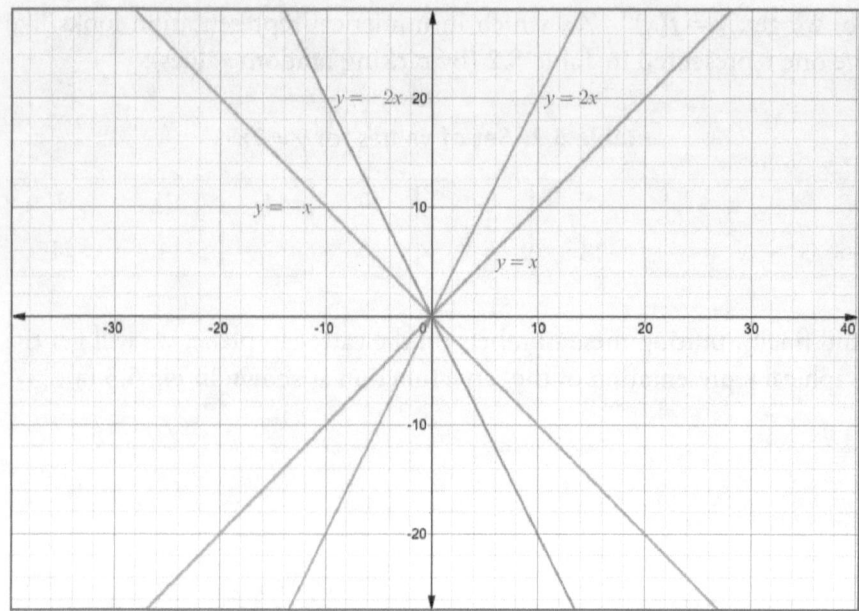

Fig. 5.4: Graph of various linear polynomials

It seems that the greater the positive value of a_1, the greater the acute angle made with the positive x-axis, and the greater the negative value of a_1, greater is the obtuse angle made with the positive x-axis (). As if controls the ability as to *how much the line will tilt or bend*. We thus call a_1 as the **slope** of line (the word slope is something we generally use in our day-to-day life).

Perhaps you can think of a thousand things around you that follow this kind of relationship. The cost C of number of items n brought at the rate of Rs. $a_1(C(n) = a_1 n)$. The speed Rajdhani Express is 130 km/hr, one can predict the distance d it will cover depending upon the time t (in hours) that it travels $(d(t) = 130t)$. May be the value of Current I flowing in a circuit depending on the voltage V applied $(I(V) = a_1 I)$. Personally, I would say, one will find linear dependence mostly in man-made situations rather than natural ones. Keep looking!

Query 5.1 We know the actual form of Linear polynomial function must be $y = f(x) = a_1 x + a_0$ (more popularly written as $y = mx + c$). Try to figure out what is the effect a_0 of on the graphical representation of this function. Pick up all sort of values and play. Feel free to use a Desmos or any graphing tool. Draw a mapping diagram as well.

5.4 Family of Even-Power Polynomial Functions

We decided to write the family of even-power polynomial functions, symbolically as:

$y = f(x) = x^n = x^{2k}$, *where k is some positive integer* (analytically)

Let's start with $k = 1$, and we are left out with $y = f(x) = x^2$. We can do the same old drama of plugging in numbers and drawing the function in graphical form as shown in Fig. 5.5.

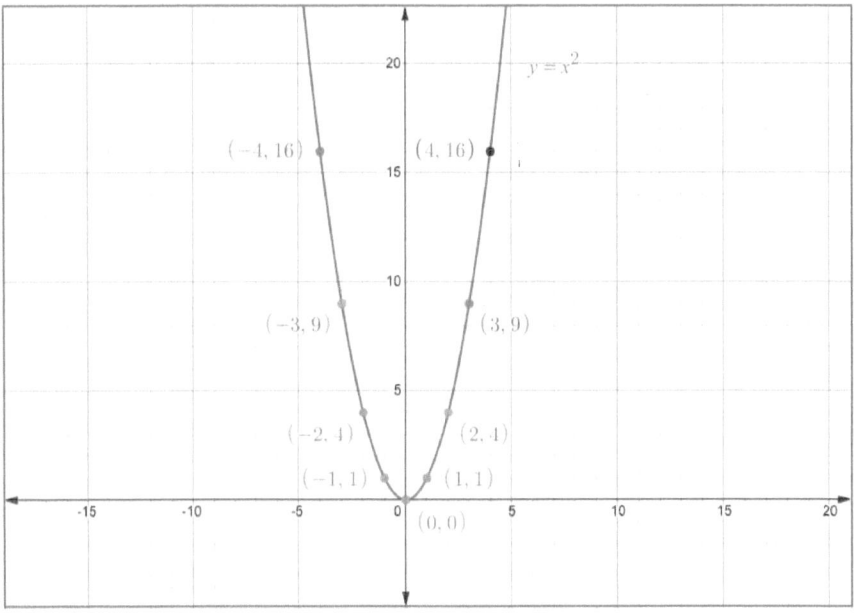

Fig. 5.5: Graph of $y = x^2$

A nice parabolic graph. Unlike the previous linear graph, this graph does not grow uniformly (does not have equal gaps in output), nor it gives us any negative values as answers. Before we move further, let's get a bit adventurous and plot a few more values of $k = 1, 2, 3, 4$ and we shall get functions like $y = f(x) = x^2, x^4, x^6, x^8$. The results in numerical form are shown in Table 5.3 and the graph is shown in Fig. 5.6.

Table 5.3: Values of various even power polynomials

x	−2	−1	−0.5	−0.1	0	0.1	0.5	1	2
$y = x^2$	4	1	0.25	0.01	0	0.01	0.25	1	4
$y = x^4$	16	1	0.0625	0.0001	0	0.0001	0.0625	1	16
$y = x^6$	64	1	0.015625	0.000001	0	0.000001	0.015625	1	64
$y = x^8$	256	1	0.00390625	0.00000001	0	0.00000001	0.00390625	1	256

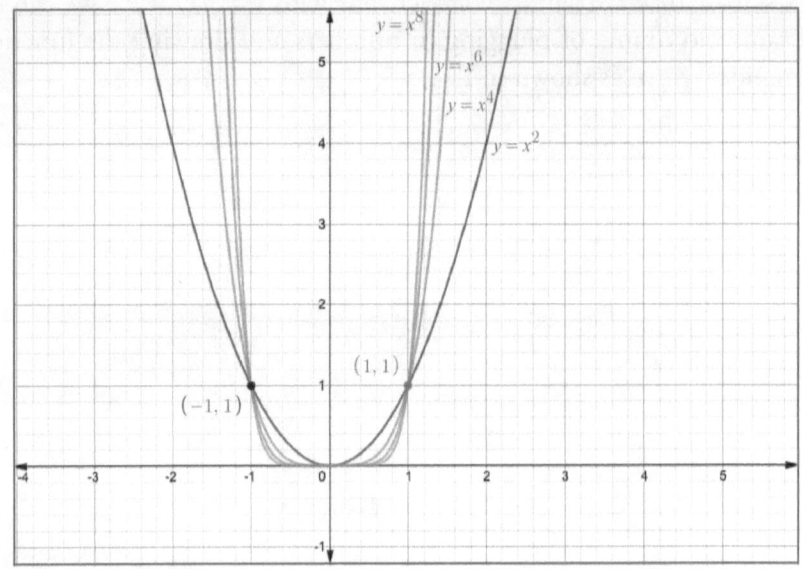

Fig. 5.6: Graph of a few even power polynomials

Some obvious things (and this holds for all even power polynomial functions): we can put any value in this function and it doesn't seem to have any problem, hence the Domain is simply set of all Real Values R. The output, however, gives only positive numbers – hence, Range is $[0, \infty)$. Notice that all graphs are symmetrical about the y-axis. Also, all graphs intersect at $(-1, 1)$ and $(1, 1)$.

Query 5.2 Analyze the graph of even power polynomial functions. What can you say about the nature of the graph from $-1 < x < 1$ (basically when the input is a decimal) versus everywhere else? (You may refer to the value table as well).

5.4.1 Da Vinci, Quadratic Functions and Mathematics of Beauty

I will now turn your focus especially on second order polynomial function and that goes without saying that they will assume the form: $f(x) = a_2x^2 + a_1x + a_0$. But before I even get there, let me talk about the guy who was one giant of a genius – *Leonardo da Vinci* (pretty fancy name).

Fig. 5.7: Leonardo da Vinci

Da Vinci showed how art and science go together to create marvels (as supposed to our dominated social thinking that students choosing art stream must be dumb and ones choosing science must be intellectual). Way back in the 15th century, he had imagined bicycles, a flying machine based on the physiology of a bat and what not! But what appeals to me most is his deeper ability to see mathematics in beauty in the form of something called the **Golden ratio.** (ϕ).

Also, called the divine ratio, it is a fairly simple concept. You take a line which is made of two smaller lines of lengths *a* and *b*. If it so happens that the ratio of length of first line (*a*) to the length of the second line (*b*) is the same as the ratio of the total length (*a* + *b*) to the length of first line (*a*). In the fancy language of algebra:

$$\phi = \frac{a+b}{a} = \frac{a}{b} \text{ (by the value is 1.618...)}$$

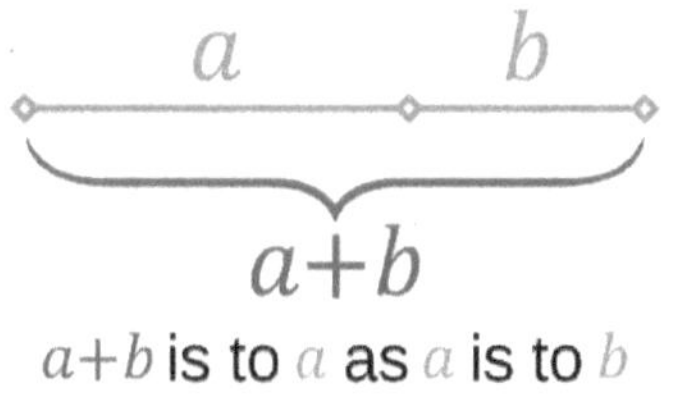

$a+b$ is to a as a is to b

Fig. 5.8: Definition of Golden Ratio

Okay, so far so good – so what do we do with this fancy ratio? It turns that out all things in nature that look "beautiful" to us, have this dimension of rectangles of *a* and *b* which measure in a way to give this golden ratio, ϕ. Look around anywhere – be it maths, geometry, design, cosmology or even life – it doesn't seem obvious but a fair number of examples are shown in Fig. 5.9

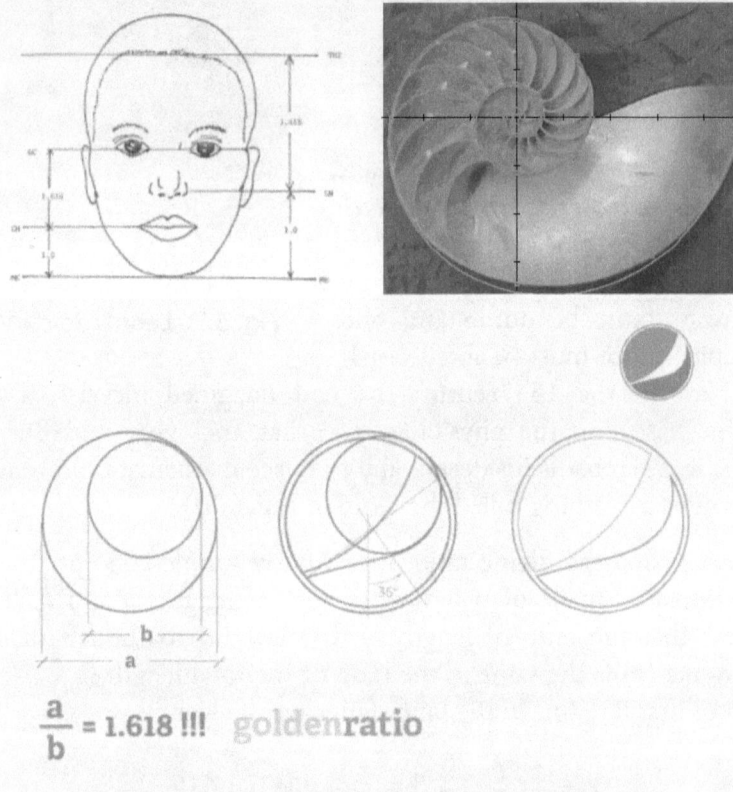

Fig. 5.9: A few instances of Golden Ratio

Hope you had your mind blown, but wait the real thing still remains. We wrote $a/b = \phi$. And we defined golden ratio as

$$\frac{a+b}{a} = \frac{a}{b}, \text{ or } \frac{a}{a} + \frac{b}{a} = \frac{a}{b} \Rightarrow 1 + \frac{b}{a} = \frac{a}{b}$$

$$\Rightarrow 1 + \frac{1}{\phi} = \phi$$

$$\Rightarrow \phi + 1 = \phi^2, \text{ or basically}$$

$$\phi^2 - \phi - 1 = 0$$

Now, look and observe. Does the equation not look familiar to you? It's like the second degree polynomial of the form: $f(x) = a_2x^2 + a_1x + a_0$, where $f(x) = 0$, $a_2 = 1$, $a_1 = -1$, $a_0 = -1$, with the only change being that now the variable is not x.

Query 5.3 Explore the relationship between Fibonacci Sequence and the Golden Ratio(ϕ).

So, somehow the idea of "beauty" in nature is hidden in a second order polynomial equation! For some reason, nature kind of knows that it has to work in rules that fit in to this particular quadratic form. But how? It is for this mysterious and many non-mysterious reasons that it becomes essential to understand how to solve quadratic equations and how they work.

A quadratic equation is of the form $a_2x^2 + a_1x + a_0 = 0$ (more popularly written as $ax^2 + bx + c = 0$). The graph undoubtedly is a parabola. Finding roots of quadratic function is to find which input variables will make the output as zero (basically where the graph crosses the x-axis).

5.4.1.1 Al-Khwarizmi's Idea of Completing the Square

al-Khwarizmi was a Persian scholar born around 780 A.D. who pioneered in al-gebra. He came up with this ingenious idea of completing the square – which by the way is something that I always prefer – convert the algebraic problem to a geometric one. I will not follow his exact method (I guess one reason being that people in that time really never knew how to deal with negative numbers and sadly we do know how to deal with them), but rather a modification to what his thinking was. Following is the illustration:

Illustration 5.2 Let's say, we wish to solve a random quadratic equation, for example: $x^2 + 6x + 16 = 0$. We can obviously do hit and trial but that won't work with every problem in hand. We use al-Khwarizmi's geometrical approach. We rewrite the equation as:

$$x^2 + 6x = 16$$

Focus on the left-hand side (LHS). There are two terms x^2 and $6x$. We can give picture to both of them. x^2 can be thought of as the area of a square whose side is x units. The term $6x$ can be broken down into two parts – $3x$ and $3x$ which can be imagined as the area of two rectangles of length 3 units and breadth x units. Fig. 5.10 (a) tells that information. We can rearrange the diagram in Fig. 5.10 (a) to look like what is shown somewhat in Fig. 5.10 (b)

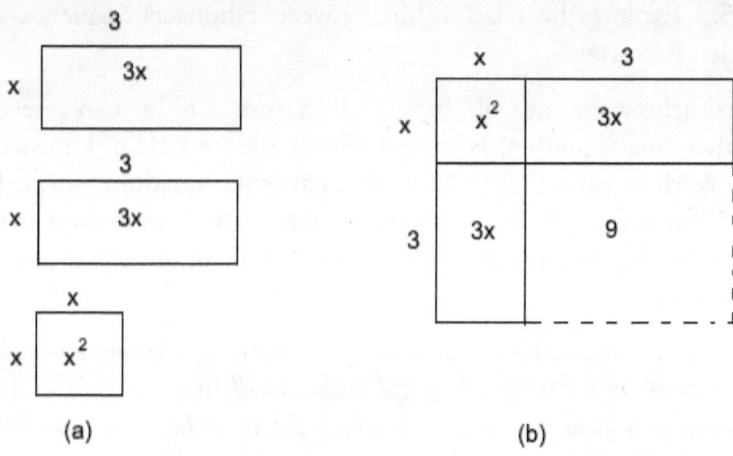

Fig. 5.10: (a) Geometrical Representation of polynomial (b) Rearrangement

We have attached the rectangles on two adjacent sides of the square of side x units. This figure is a bit incomplete. We get something close to a square, because the two sides measure $x + 3$ units each. The dotted line will show that finished figure that we need to "complete the square" and yes, the area need more is 9 units. This is what we do to the quadratic equation:

$x^2 + 6x + 9 = 16 + 9$ (adding 9 units area to both sides to maintain the balance)

Now, $x^2 + 6x + 9$ is basically the total area of the square that has side length of $x + 3$ units. But we also know that area of square is just square of side length and hence,

$(x + 3)^2 = 25$, which is more sensible question now – that literally says that my square should have an area of 25 units. And now guessing the answer shouldn't be a big deal! x should simply be 2 units (as 2 + 3 = 5 and a square with side length 5 will have an area of 25). Well, that's the obvious guess, what's not so obvious is that x could also be -8 (as $-8 + 3 = -5$, the square of which is also 25) but that would not have made any sense to al-Khwarizmi because side lengths being negative was nonsensical to him. But that doesn't mean it has no meaning.

Learning to complete the square is one of the most important skills you'll possess in learning calculus. It may not seem like now, but hold on – you'll see.

Query 5.4 It is quite easy to see how the most common formula $(a + b)^2 = a^2 + 2ab + b^2$ fits into the picture of completing the whole square. Can you come up with a picture of $(a - b)^2 = a^2 - 2ab + b^2$ also?

5.4.1.2 Brahmagupta's Discriminant Formula

I mentioned Brahmagupta earlier in Chapter 1. His Brāhmasphuṭasiddhānta gives key ideas about algebra. One of them being the discriminant formula. Although, historically Brahmagupta was born well before al-Khawarizmi, he came up with a rather ingenious idea of dealing with quadratic equations. Again, I use modifications to his actual formula in order to fit the modern context (keeping in mind, that people in those days actually had problems with negative and irrational numbers). Perhaps, it would have occurred to him that this method of drawing and completing squares is too tiresome and no-one would want to do it again and again. So, he comes up with a general idea. A general quadratic equation has the formula:

Fig. 5.11: Brahmagupta

$ax^2 + bx + c = 0$ (popular form), where a, b, c can be any numbers (or say squares and rectangles of any length).

We divide by a throughout and push the c to the other side to make it look like this:

$$x^2 + \frac{b}{a}x = -\frac{c}{a}$$

This kind of looks familiar, isn't it. See, for yourself:
$x^2 + 2\frac{b}{2a}x = -\frac{c}{a}$, will look more obvious with Fig. 5.12.

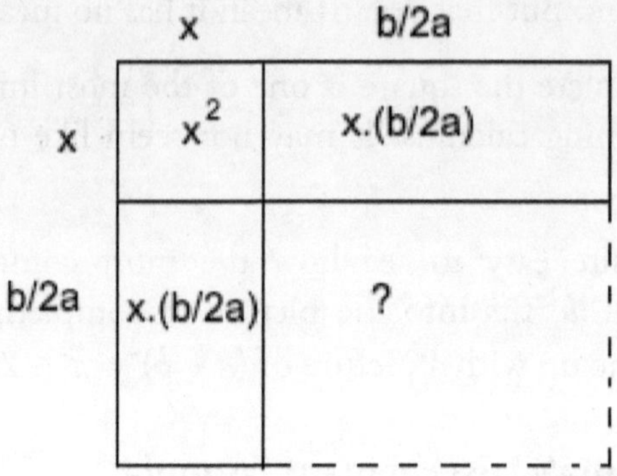

	x	b/2a
x	x^2	x.(b/2a)
b/2a	x.(b/2a)	?

Fig. 5.12: Geometrical representation of completing the squares

So, clearly the missing element to complete the whole square is $\left(\frac{b}{2a}\right)^2$, which we add to both sides:

$$x^2 + 2\frac{b}{2a}x + \left(\frac{b}{2a}\right)^2 = -\frac{c}{a} + \left(\frac{b}{2a}\right)^2,$$

$$\Rightarrow \left(x + \frac{b}{2a}\right)^2 = -\frac{4ac}{4a^2} + \frac{b^2}{4a^2},$$

$$\Rightarrow \left(x + \frac{b}{2a}\right)^2 = \frac{b^2 - 4ac}{4a^2},$$

$$\Rightarrow \left(x + \frac{b}{2a}\right) = \pm\sqrt{\frac{b^2 - 4ac}{4a^2}}$$ (essentially because we have two roots)

$$\Rightarrow x = \frac{-b}{2a} \pm \frac{\sqrt{b^2 - 4ac}}{\sqrt{4a^2}}, \text{ or finally}$$

$$x = \frac{-b \pm \sqrt{b^2 - 4ac}}{2a}$$

And there you go! If you happen to know a, b and c, you can literally plug it into the formula above called discriminant formula which gives you answers/roots directly! One should also clearly understand the term under the root here, namely, $b^2 - 4ac$ is what we call determinant. If is less than zero, we shall have no real solutions; if it is equal to zero, we will two equal real solutions and it is greater than zero, we shall have two distinct real solutions.

Query 5.5 Can you now use Brahmagupta's discriminant formula to figure out the value of the golden ratio which is given by the following quadratic form:

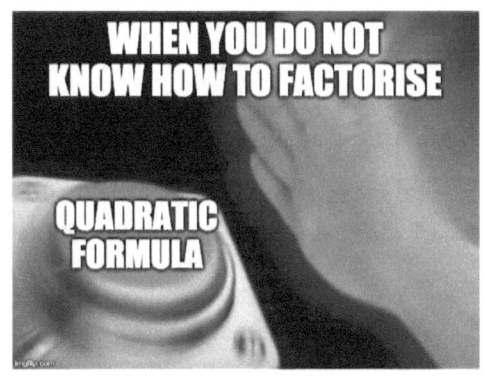

Meme 5.2: Use discriminant formula

$$\phi^2 - \phi - 1 = 0$$

Illustration 5.3: Let us discuss the relationship between the product and sum of roots of quadratic equations and its coefficients. As an example, let us take the quadratic equation: $2x^2 + 5x + 3 = 0$ (General form being $ax^2 + bx + c = 0$). You can find the roots (we shall denote the roots as α and β) of this equation completing the square or factorising – your wish. We will use factoring here. We can write the same equation as:

$$2x^2 + 2x + 3x + 3 = 0$$

$$\Rightarrow 2x(x+1) + 3(x+1) = 0,$$

$$\Rightarrow (x+1)(2x+3) = 0$$

Or, $x = -1 \, or \, x = -\dfrac{3}{2}$

Now, we find what the product and sum of these roots are. Clearly, $\alpha + \beta = -1 + (-3/2) = -5/2$ and $\alpha\beta = (-1)(-3/2) = 3/2$. You should notice perhaps some patterns in these results. In fact,

$$\alpha + \beta = -\frac{coefficient\ of\ x}{coefficient\ of\ x^2} = -\frac{b}{a}, \text{ and}$$

$$\alpha\beta = \frac{constant\ term}{coefficient\ of\ x^2} = \frac{c}{a}$$

Could you try proving this observation? In fact, this relationship between product and sum of roots with coefficients exists for all polynomials given by Vieta's formulas (remember somebody of that name from starting of this chapter?) which says:

If there is a polynomial of the form $a_n x^n + a_{(n-1)} x^{(n-1)} + a_{(n-2)} x^{(n-2)} + ... + a_2 x^2 + a_1 x^1 + a_2$ which has n number of zeros given by $a_1, a_2, a_3, ... a_n$, then:

$$\alpha_1 + \alpha_2 + \alpha_3 + ... + \alpha_n = -\frac{a_{n-1}}{a_n} \text{ and}$$

$$\alpha_1 . \alpha_2 . \alpha_3 ... \alpha_n = (-1)^n \frac{a_0}{a_n}$$

Query 5.6 What is the minimum amount of information needed to uniquely determine a straight line or a parabola?

5.5 Family of Odd Power Functions

We decided to write the family of odd-power polynomial functions, symbolically as:

$y = f(x) = x^n = x^{2k+1}$, where k is some positive integer (analytically)

Let's start with $k = 1$, and we are left out with $y = f(x) = x^3$. We can do the same old drama of plugging in numbers and writing the function in graphical form as shown in Fig. 5.13.

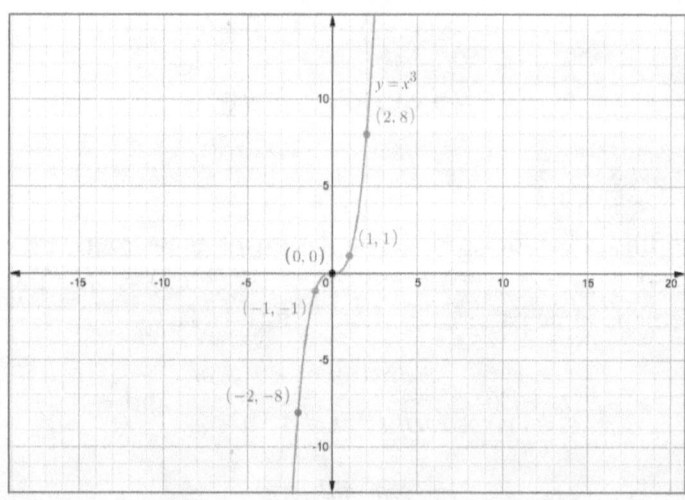

Fig. 5.13: Graph of $y = x^3$

Pretty, weird graph! One may find it difficult to see any line of symmetry here! But keep looking! Notice that this graph, unlike the even-power polynomial functions does give us negative values as answer. As usual, let's get a bit adventurous and plot a few more values of $k = 1, 2, 3$ and we shall get functions like $y = f(x) = x^3, x^5, x^7$. Some of the values are shown in Table 5.4 and the graphs are shown in Fig. 5.14.

Table 5.4: Values of various odd power polynomials

x	−2	−1	−0.5	−0.1	0	0.1	0.5	1	2
$y = x^3$	−8	−1	−0.125	−0.001	0	0.001	0.125	1	8
$y = x^5$	−32	−1	−0.03125	−0.00001	0	0.00001	0.03125	1	32
$y = x^7$	−128	−1	−0.0078125	−0.0000001	0	0.0000001	0.0078125	1	128

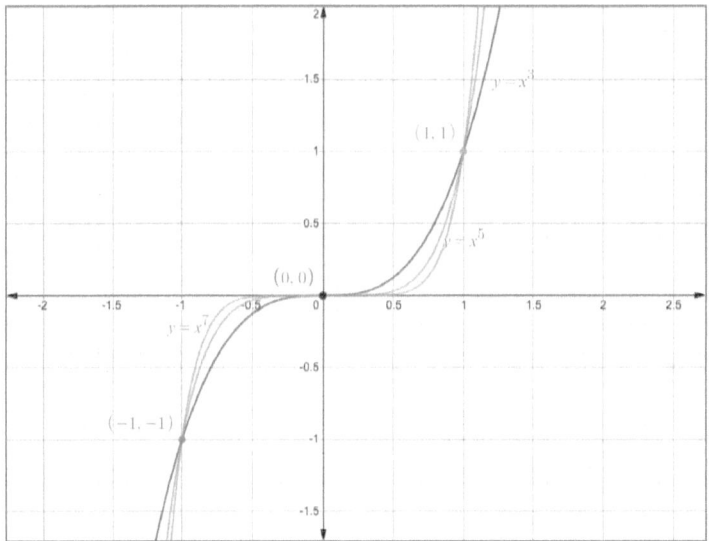

Fig. 5.14: Graph of some odd power polynomials

Some obvious things (and this holds for all odd power polynomial functions): we can put any value in this function and it doesn't seem to have any problem, hence the Domain is simply set of all Real Values R. The output unlike the even power polynomials does give all values possible, hence, the range is also R. Notice that all graphs intersect at $(−1, −1)$ and $(1, 1)$.

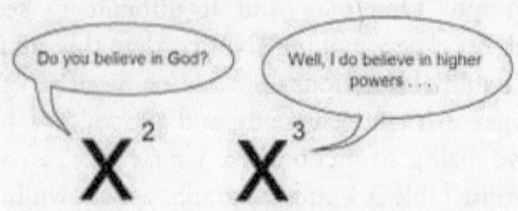

Meme 5.3: "Bhagwan mei maante ho?"

Query 5.7 Analyze the graphs of odd power polynomial functions. What can you say about the nature of the graph from $-1 < x < 1$ (basically when the input is a decimal) versus everywhere else? Also, verify that the line $y = x$ is the line of symmetry. (You may refer to the value table as well).

Query 5.8 Use Vieta's formulas to find the relationship between sum and product and the coefficients of a third order polynomial of the form $ax^3 + bx^2 + cx + d = 0$

Query 5.9 Can you derive a general formula for solving $ax^3 + bx^2 + cx + d = 0$?

Illustration 5.4: A little note about division of polynomials here. We start this illustration by Remainder theorem. It says that if we divide a polynomial $p(x)$ by a linear polynomial $(x - a)$, then the remainder happens to be simply $p(a)$. To verify this, let us take a polynomial $p(x) = x^3 + 3x^2 + 4x - 12$ which is divided by $(x - 2)$. In order to find the remainder, we must plug 2 in the polynomial, i.e.

$$p(2) = 2^3 + 3(2)^2 - 4(2) - 12 = 8 + 12 - 8 - 12 = 0.$$

Hence, we know already that these two polynomials are completely divisible as the remainder is zero (this special case is also what we call factor theorem).

Now suppose, we want to know the quotient. Of course, we have the long division method. But I would propose you to follow a short mental trick. Let's try to divide $p(x) = x^3 + 3x^2 + 4x - 12$ by $(x - 2)$ Since the polynomial $p(x)$ is of third degree, we will write $(x - 2)$ three times with a few dashes like this:

$$\_(x-2) + \_(x-2) + \_(x-2)$$

The goal will be to create the original polynomial in this notation. The first term is x^3, therefore if we write x^2 in the first dash, we get:

$$x^2(x-2)+\_(x-2)+\_(x-2)$$

x^2 multiplied by $(x-2)$ does gives us x^3 but it also gives us a $-2x^2$, but do we want that? No! We want a $3x^2$, so to counteract that extra term, we put in the next dash, and we get:

$$x^2(x-2)+5x(x-2)+\_(x-2)$$

Now, we get $5x^2$ (which counteracts with $-2x^2$ to give us $3x^2$ as we wanted) and a $-10x$. But again, do we want a $-10x$? Thus, in order to kill that, we add 6 in that last dash.

$$x^2(x-2)+5x(x-2)+6(x-2)$$

Which gives a $6x$ (which counteracts with $-10x$ to give us $4x$) and -12, which we wanted anyway! Now, we pull the common terms out and we get:

$$(x^2+5x+6)(x-2)$$

And that's all! We have got the quotient as x^2+5x+6. I know it looks really big in writing, but it is just a single line problem when you do it in head and paper.

Query 5.10 Usually while factoring a polynomial like $x^2+5x+6=0$, we usually prefer to have 0 in the RHS. Why is that so? Is it just a convention or has some benefits?

Illustration 5.5 Descartes' Rule of Signs. Here is another tool in your box that can be quite handy at times. This one lets you tell how many real roots a polynomial has. The idea is quite simple indeed.

For example, let us take a polynomial of fifth degree, $p(x)=5x^5+4x^4-3x^3+2x^2+3x-6$. Now look at the signs of the coefficients – they change from +5 to +4 to −3 to +2 to +3 to −6. We count the number of times the sign has "actually" (i.e. from + to − or − to +). In this case, three times (check!). We can now say that this polynomial has a maximum of 3 real roots. And that's it!

Meme 5.4: Sadness

5.6 Closure

We have now come end to our first branch of functions and first sub-branch of algebraic functions. Looking around you, you'll find polynomials being used without even being aware of it. Almost all dependencies in nature that linear, quadratic or of any power are found all the time. A taxi driver uses linear polynomials to charge you money. Dependence of electric current on voltage is a linear one. A ball thrown at someone follows the rules of quadratic form and so does energy stored in a spring. Business companies make polynomial models to maximize profits. Even the construction of roller coaster finds its origin in polynomials (particularly sixth degree polynomials)! While I do agree that this chapter requires a lot of symbolic stuff to get through in mind – but trust me the end result is understanding almost all of the things that go around you – the mathematical models beneath them.

Exercises

1. Prove that the value of $11 - 4x - x^2$ can never be greater than 11. (Use graph or completing the squares)

2. Tushar throws a ball in the air towards his friend. The horizontal motion of ball is modeled by $x = 0.707t$ and the vertical motion is modeled by the equation $x = 0.707t - 4.9t^2$. Find the equation of the path (y as a function of x). Which kind of polynomial is it? Use method of completing the squares to tell the maximum height the ball reaches.

3. Most of us watch television (I strictly advise you not to). The aspect ratio of a flatscreen T.V. is the ratio of its width to its height. Let's say that you buy a 32 inch television (ever wondered what is the meaning of that? Its referring to the length of the diagonal of television) which has an aspect ratio of 16:9. Find the area of the flat screen TV.

4. Suppose a local newspaper in your town has about 50,000 readers when it charges 5 rupees per copy of paper. Data reveals that for each 1 rupee increase in the price, the paper loses 2000 readers. How much should the newspaper increase the maximum profit? What is the maximum profit?

5. You are spraying water from your garden pipe to the plants. The height y, in meters, of the water sprayed from the pipe, is given by $y = -x^2 + 6x + 7$. How high does the water go? How far from you horizontally does water go to its maximum height?

6. Find the range of values of β for which the line $y = -x + 2\beta - 5$ will be secant to the curve $y = x^2 + 3\beta x + 5$.

7. Find all possible solutions for x if, $x^3 - 3x + 2$ is a prime number.

8. The points $(-1, -1)$ and $(-1, -12)$ lie in the graph of the function of the function $\mu x^2 + \lambda x^3$. Find the values of μ and λ.

9. Let us take a second – degree polynomial, namely, $p(x) = ax^2 + bx + c$. Under what conditions will the expression always give positive numbers?

10. Find the sum of the roots of the polynomial $p(x) = 2^{333x-2} + 2^{111x+2} - 2^{222x+1} - 1 = 0$ having three real roots.

11. Suppose that the roots of $x^2 + 4x^2 + 5x - 13 = 0$ are and α, β and γ that the roots of $x^3 + ax^2 + bx + c = 0$ are $\alpha + \beta$, $\beta + \gamma$ and $\gamma + \alpha$. Find the value of c.

12. In physics, we use Van der Waals equation to relate various properties of real gases like Volume(V), Temperature(T) and Pressure(P). It is written as:

$$V^3 - \frac{1}{3}(1+\frac{8T}{P})V^2 + \frac{3}{P}V - \frac{1}{P} = 0$$

where volume is measured in dm^3, temperature is measured in Kelvin and pressure is measured in atmospheres. In a situation called the critical temperature which happens at T = 1K and P = 1atm, find the volume using the Van der Waals equation.

13. Given that $2x^2 + x + 5$ is a factor of $2x^3 + \gamma x^2 + \delta x + 15$. Find the values of γ and δ.

14. Find all points where graph of $y = 3x - 1$, $y = x^2 + 1$ and $y = x^3 + 4$ intersect. Do they all ever intersect together at one point?

15. Find the range of the following: a) $x^2 + 5$ b) $x^2 + 3x + 2$

16. The Yuva Bharti Krirangan Stadium in Kolkata is the second largest football stadium in the world having a capacity of 85000 people. With an attendance of 80,000 people, the tickets were priced at 1000 rupees. A newspaper reports that for every 100 rupees decrease in ticket price, the audience increases by 700 people. Find a function that models revenue of the stadium in terms of ticket price.

17. Consider all lines that meet the graph of polynomial $p(x) = 2x^4 + 7x^3 + 3x - 5$ at the points (x_1,y_1), (x_2,y_2), (x_3,y_3) and (x_4,y_4). Show that $x_1 + x_2 + x_3 + x_4$ does not depend on what line intersects and find its value. (Hint: Use Vieta's theorems)

6 May the Force Be with You: Rational and Irrational Functions

"One of the most amazing things about mathematics is the people who do math aren't usually interested in application, because mathematics itself is truly a beautiful art form. It's structures and patterns, and that's what we love, and that's what we get off on."

— Danica McKellar

Hooke and Newton are both well-known fellows. Hooke, born in 1635, in the same country as Newton was the discoverer of *cell – the fundamental unit of life*, established that light is a wave and forget all that – the watches that you wear and see all around – they work because of the spring balances developed by him. You may assume a man of that stature must have had a cool life. Well, quite the opposite. Hooke went downright into the dustbin of history of science. Not a single book celebrates him as he deserves to be celebrated. But, why?

6.1 Hooke, Newton, Universe and Rational Functions

Hooke was bubbling with ideas all the time. So much so, that in spite of all the fame, he wanted credit for every idea that came along at that time. And to be fair, his ideas were way ahead of his time. He hypothesized about fossils and assumed that they would be dead remains of animals that became extinct because of some sort of natural mass destruction (dinosaurs, sad life). But that idea was straight up rejected by Church as it would be against the idea of perfect world made by God (as I have mentioned in the previous chapter). One of the crucial problems was about how planets move about in space? Exactly in which shape? Is there a universal force pulling all planets together? Big questions! Kepler did come up with experimental data but there was no mathematics describing it. Halley (after whom the famous Halley's comet is named) went through Kepler's data and had guessed a "dependence/relation" between the forces (F) on planets and how far they are (r) from sun ($F = f(r)$).

But science doesn't work on guesses, does it? However, it was a neat guess. Kepler, based on his experimental data, had said that planets move in elliptical paths – however, he never explained why do they do so? If the "dependence/relation" that Halley had guessed was correct, then that would actually lead to planets moving in elliptical paths. But obviously, he needed someone to prove it.

Halley and Hooke met for coffee (in 1684) along with a third scientist named Christopher Wren where Halley put forth this problem. Wren immediately understood the importance of this and announced a huge cash prize for whoever proves that the "dependence/relation" guessed by Halley actually makes the planet go in elliptical paths. And obviously, Hooke said he can prove it. He would go crazy for the rest of his lifetime trying to defend his intelligence and proving it – but Halley never received any proof from him. Later that magic happened when Halley met Newton. I will write it here exactly as described by De Moivre:

"In 1684 Dr. Halley came to visit him at Cambridge. After they had been some time together, the Dr. asked him what he thought the curve would be that would be described by the planets supposing the force of attraction towards the sun to be reciprocal to the square of their distance from it. Sir Isaac replied immediately that it would be an ellipse. The Doctor, struck with joy and amazement, asked him how he knew it. Why, saith he, I have calculated it. Whereupon Dr. Halley asked him for his calculation without any further delay. Sir Isaac looked among his papers but could not find it, but he promised him to renew it and then to send it to him..."

And boom, that is how the publication of the greatest text in the history of science "Principia" by Issac Newton happened where he finally proved that if that *"dependence/relation"* guessed by Halley is right, then planets do move in ellipses! After shaking the entire scientific world with his new vision of world – Newton did make Hooke look like a loser upon which Hooke alleged that Newton had copied his work. Newton, being the introvert and proud person as he was, never really let the allegations go away.

According to legend, after Hooke's death – when Newton became head of Royal Society, he made sure all portraits of Hooke was burned and till date none of his portraits about he looked like exists. All this life-long drama, all this jealousy, hatred because of one *"dependence/ relation"* given by Halley. But what was the dependence that he had discovered?

He had said that, that the sum of forces (F) acting on a planet depends on the inverse of the second power of its distance from the sun $\left(\frac{1}{r^2}\right)$, or

$$F(r) \propto \frac{1}{r^2}$$

Which belongs to a new branch of algebraic functions called the *"rational functions"* (So much irrational behavior for so called rational functions!).

We formally define rational functions as follows:

If $p(x) = a_n x^n + a_{n-1} x^{n-1} + a_{n-2} x^{n-2} + \ldots + a_2 x^2 + a_1 x^1 + a_0$ and $q(x) = b_m x^m + b_{m-1} x^{m-1} + b_{m-2} x^{m-2} + \ldots + b_2 x^2 + b_1 x^1 + b_0$ are some polynomial functions, $a_n, a_{n-1} \ldots a_1, a_0$ and $b_m, b_{m-1} \ldots b_1, b_0$ are some real numbers, being positive numbers (degrees of $p(x)$ and $q(x)$ respectively), then a rational function is defined by,

$$f(x) = \frac{p(x)}{q(x)} = \frac{a_n x^n + a_{n-1} x^{n-1} + a_{n-2} x^{n-2} + \ldots + a_2 x^2 + a_1 x^1 + a_0}{b_m x^m + b_{m-1} x^{m-1} + b_{m-2} x^{m-2} + \ldots + b_2 x^2 + b_1 x^1 + b_0}$$

Note that the degrees $p(x)$ of $q(x)$ and need not be same.

Already looks pretty intense, does it? Well, just looks scary – in a nutshell, it is literally a division of two polynomial functions. That's all. Now, if we set $p(x) = 1$ (constant polynomial) and $q(x) = x^2$ (second degree polynomial), we get the rational function,

$$f(x) = \frac{p(x)}{q(x)} = \frac{1}{x^2}$$

Meme 6.1: Newton got no chill

which is kind of the "dependence/relation" that Halley discovered. We start with this seemingly easy form of rational function where the numerator is set to 1. A more general form is:

$$f(x) = \frac{p(x)}{q(x)} = \frac{1}{x^n}$$ (also called as *"inverse variation"*)

where n is a positive integer. As usual, n can be either even or odd. If $n = 2k$, it's even and if $n = 2k + 1$, where k is any positive integer. Next up, we will explore this inverse variation rational function in two sections.

Query 6.1 Can a function be rational even though it has only irrational values (like π etc.)?

6.2 Rational Functions of Type $\frac{1}{x^n}$, $n = 2k + 1$

Naturally, these rational functions have an odd-power polynomial in the denominator. If I may remind you, all polynomials don't really have an issue with what goes in as an input to them i.e. their domain is simply the set R of all real numbers, always. Is it the same with rational functions given that we are dividing two polynomial functions together? Remember, "division"! If we put k = 0, we get the simplest type of rational function:

$$y = f(x) = \frac{1}{x}$$

We will try to analyze this supposedly simple graph by the following illustration:

Illustration 6.1 Before even going into technical details, let's use common sense here. If the dependence of y on x is $y = \frac{1}{x}$, then it becomes clear that the larger the x, the smaller the value of y. In fact, let's draw a table here (to numerically represent this function) as shown in Table 6.1

Table 6.1: Set of values for $y = f(x) = \dfrac{1}{x}$

x	−100	−50	−25	−1	−0.5	0.5	1	25	50	100
$y = f(x)$	−0.01	−0.02	−0.04	−1	−2	2	1	0.04	0.02	0.01

Notice the trend? If x is "very big negative number", then y becomes a "very small negative number"; similarly if x is "very big positive number", then y becomes a "very small positive number". But what happens if x is very close to zero or in fact, it is zero! If it doesn't strike you now, go straight to the very first section of the very first chapter this book and you'll know why I wrote it. Remember from there that:

$$\frac{1}{a \ very \ very \ small \ number} = a \ very \ very \ big \ number, \text{ symbolically}$$

written as $\dfrac{1}{0} = \infty$. And now we are talking. Furthermore, and I think it should make sense:

$$\frac{1}{a \ very \ very \ small \ negative \ number(0)} = a \ very \ very \ big \ negative \ number$$

$(-\infty)$ and $\dfrac{1}{a \ very \ very \ small \ positive \ number(0)} = a \ very \ very \ big \ positive$ number $(+\infty)$

Stressing again and again, that infinity is just a concept as I had mentioned in Chapter 1. It means that as we put values of x closer to 0 from right of number line (very small positive number), the y shoots up to $+\infty$ and as we put values of x closer to 0 from left of number line (very small negative number), the y shoots up to $-\infty$. We can't actually divide by zero, and hence don't know what happens when x actually becomes zero.

Another point before we finally graph this function, what happens when becomes like really big? Should be quite easy to predict now.

$$\frac{1}{a \ very \ very \ big \ negative \ number(-\infty)} = a \ very \ very \ small \ number \ (0) \quad \text{and}$$

$$\frac{1}{a \ very \ very \ big \ positive \ number(+\infty)} = a \ very \ very \ small \ number \ (0)$$

Or basically, when x gets like real big from both sides (positive and negative), y simply shrinks down to zero. Using all the observations, we finally graph it as shown in Fig. 6.1

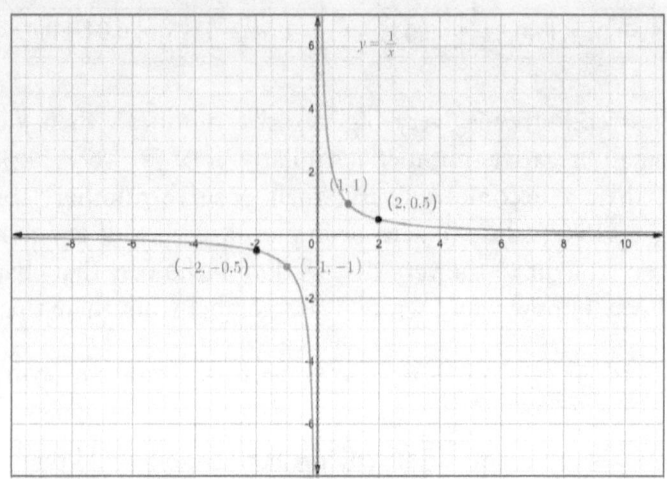

Fig. 6.1: Graph of the function $y = f(x) = \dfrac{1}{x}$

I do agree that it is one weird graph after all! The input allows itself to be all values except 0 (because then it goes all crazy and doesn't know what to do!) and hence domain is R − {0}. The output also does the same, if you observe and hence the Range is R − {0}.

So here is one key rule to keep in mind: the moment you see a rational function, you make sure to remove all such values from domain that make the polynomial function in the denominator go 0 (and hence making the output go crazy).

Exemplum 6.1: Find the domain and range of the function $f(x) = \dfrac{1}{2x-3}$.

Solution: Of course, as discussed you don't want the denominator to go zero. Symbolically,

$$2x - 3 \neq 0, \text{ or simply, } 2x \neq 3 \text{ and } x \neq \frac{3}{2}.$$ Therefore, Domain is R − {3/2}. Range is essentially everything except 0. Try to see it for yourself.

If we come back to investigating $f(x) = \dfrac{1}{x^n}$, where n is an odd number, we can continue getting higher powers as n = 3, 5, 7…and graph them all in the same way by plugging in a few numbers (both positive and negative). We collectively draw them together as shown in Fig. 6.2.

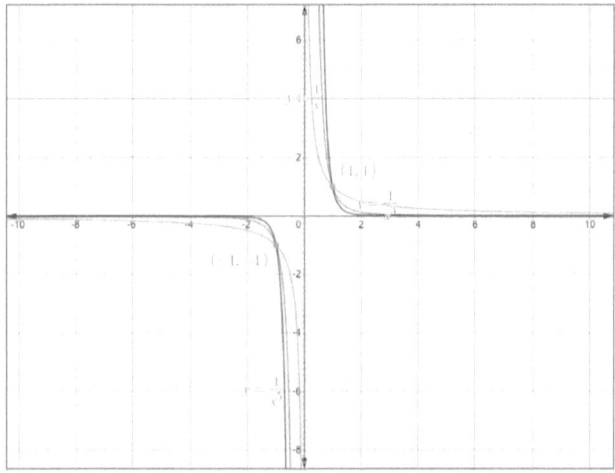

Fig. 6.2: Graph of the function $y = f(x) = \dfrac{1}{x^n}$, where n is odd

6.3 Rational Functions of Type $\dfrac{1}{x^n}$, n = 2k

Naturally, these rational functions have an even-power polynomial in the denominator. If we put k = 1, we get the rational function that created havoc between Hooke and Newton:

$$y = f(x) = \dfrac{1}{x^2}$$

We do have certain expectations here in comparison with Section 6.2. First, the value of y will still decrease with x, but of course, more *rapidly* now. Next thing, that all values that we get in output will be positive because of the degree 2 in the denominator. Also, as x goes near to zero from any side, the y value will always go towards +∞ and never towards −∞ (again, because of the degree 2 polynomial which always gives positive answers). With a few values in hand, we get a graph like the one shown in Fig. 6.3.

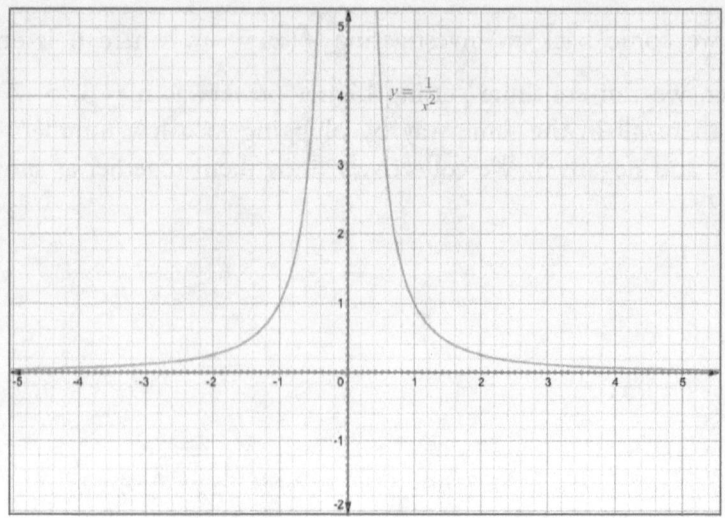

Fig. 6.3: Graph of the function $y = f(x) = \dfrac{1}{x^2}$

Again, the Domain happens to be R −{0} but range is only (0, ∞)

Query 6.2: Draw the graphs for the family of functions $y = f(x) = \dfrac{1}{x^n}$ when n is an even integer. Also write their domain and range.

Meme 6.2: Don't:(

6.4 First Encounter with Mathematical Aliens: The Asymptotes

We have been graphing functions for quite some time now. Notice something weird in this chapter? The last two sections. Heck! Go back and see what happens to the graph of both $\frac{1}{x}$ and $\frac{1}{x^2}$ near the x-axis and y-axis. They badly want to touch it, but they just can't. I mean they can be as close as possible – but sadly, they cannot ever, ever touch it. Why so? Because as mentioned again and again in Chapter 1, infinity is nothing but a concept – a concept we badly wished was actually a number that we could deal with – well, not for now at least. The x-axis and y-axis in these examples, are what we call as *asymptotes* (the x-axis being the horizontal asymptote and the y-axis being the vertical asymptote). More formally,

An asymptote is a line (or sometimes even a curve) that the graph of a function approaches extremely closely, but never touching it.

Of course, the graphing has become interesting now. The whole point of having rational functions was to introduce the asymptotes. Now, asymptotes (if they are straight lines) are of three kinds (shown in Fig. 6.4):

a. Horizontal asymptote

b. Vertical asymptote

c. Slant asymptote

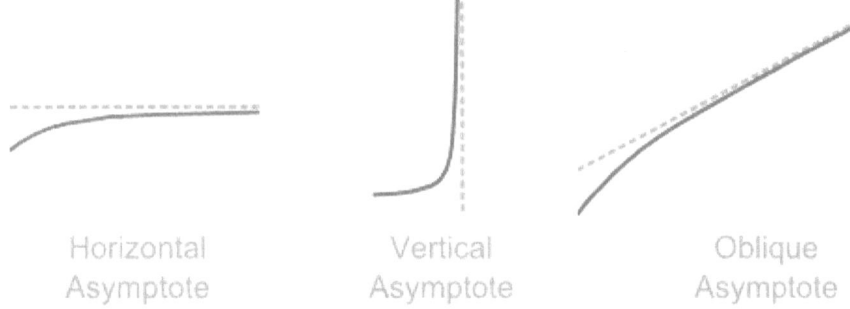

Fig. 6.4: Different kinds of asymptotes

Let's go back to the definition of Rational functions, namely:

$$f(x) = \frac{p(x)}{q(x)} = \frac{a_n x^n + a_{n-1} x^{n-1} + a_{n-2} x^{n-2} + \ldots + a_2 x^2 + a_1 x^1 + a_0}{b_m x^m + b_{m-1} x^{m-1} + b_{m-2} x^{m-2} + \ldots + b_2 x^2 + b_1 x^1 + b_0}$$

Where the degree of the numerator is n and the degree of the denominator is m. Now all this asymptote business depends a lot upon degrees of numerator and denominator. This is why, just to make matters simple, I started out by putting $p(x) = 1$, or the degree of numerator the lowest that was possible. These things look quite alien, don't they? But don't worry, with time and practice – they'll look friendly. Never easy in the future, mind you – but yes, friendly (Notice that Illustration 6.1 was a very subtle intro to asymptotes).

As we have encountered before, graph of a single function can have multiple asymptotes. The key is that we look for them one by one.

6.5 Climbing Higher and Higher: Vertical Asymptotes

Picture this: we got a vertical line-upright! Our graph wants to get as close to it as it wants. Our reference vertical line is the y-axis itself. All other vertical lines are parallel images of it. So if our y-values (or the output of the function) become really really huge but still can't touch a vertical line, that's what we would call a vertical asymptote. Isn't it? The key is this: how do you get to make a y-value infinitely huge (or simply say ∞) in $y = f(x) = \dfrac{p(x)}{q(x)}$?

Well of course, by making the denominator (in this case $q(x)$) an extremely small number (simply say close or 0 or 0 itself).

Illustration 6.2 Let's focus ourselves completely on the vertical asymptotes of the graph of the following function:

$$f(x) = \frac{1}{(x^2 - 1)(x - 3)}$$

Now obviously this graph has a lot more to it than just vertical asymptotes, it's got some crazy flips and curves probably a horizontal asymptote, but for the time being – let's concentrate on the matter at hand. Luckily, vertical asymptote has nothing to do with the degrees of numerator and denominator. We only have to focus on what makes y ($f(x)$) go ∞ or denominator go 0. Breaking the denominator into its factors, we get:

$$f(x) = \frac{1}{(x-1)(x+1)(x-3)}$$

As you can observe, the denominator can go zero in three ways: by putting $x = 1$, $x = -1$ and $x = 3$. So that's all we draw these vertical lines and we can be sure that the graph will try to approach them, but never really touch it. You can obviously put numbers in place of x, and get a few values as output to plot the graph, but specifically for the vertical asymptotes, here is the graph generated by the computer as shown in Fig. 6.5.

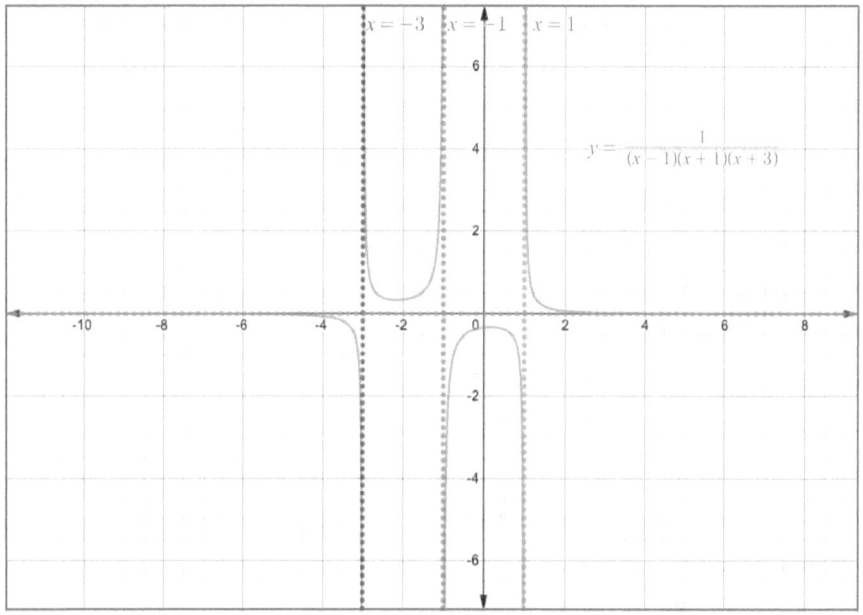

Fig. 6.5: Graph of the function $y = f(x) = \dfrac{1}{(x-1)(x+1)(x-3)}$

6.6 Getting Close to the Ground: Horizontal Asymptotes

Now that you are familiar with vertical asymptotes, we will move to a slightly more complicated case. The ideal/reference horizontal line is just the x-axis, isn't it? All other horizontal lines are just parallel images of it. The key thing to note is that as the x-values become larger and larger they approach a horizontal line, but never really touch it. However, this does depend on the degree of the numerators and denominators.

From Section 6.4, the degree of the numerator is n and the degree of the denominator m. A horizontal asymptote will occur if the degree of the numerator is less than or equal to the degree of the denominator (i.e. $n \leq m$). Hence, the two cases:

Case 1: When $n < m$, in this case the horizontal asymptote is always the $y = 0$ (or simply say the x-axis).

Case 2: When $n = m$, in this case the horizontal asymptote is the line $y = \dfrac{a_n}{b_m}$ (or simply say the ratio of leading coefficient of numerator and the leading coefficient of the denominator).

Quite simple! Of course, there is why and how involved here. I have not given you the explanation here. But let's have some suspense. The entire calculus is waiting for us to answer this question.

Exemplum 6.2: Find the horizontal asymptote in the graph of function given in Illustration 6.2.

Solution: The given function is:

$$f(x) = \frac{1}{(x^2 - 1)(x - 3)}$$

Clearly, the numerator has a degree less than the denominator and hence the graph will have a horizontal asymptote at $y = 0$ (or the x-axis). The graph drawn in Fig. 6.5 confirms this fact.

Exemplum 6.3: Find the horizontal asymptote in the graph of the function:

$$f(x) = \frac{3x^2 + 1}{(x^2 + 1)}$$

Solution: First things first, the graph will have no vertical asymptotes, because nothing makes the denominator go zero. The numerator and denominator both have a degree of 2. The leading coefficient of numerator is 3 and that of the denominator is 1. Hence, the horizontal asymptote is $y = 3/1 = 3$. We have plotted the function to verify this fact as shown in Fig. 6.6.

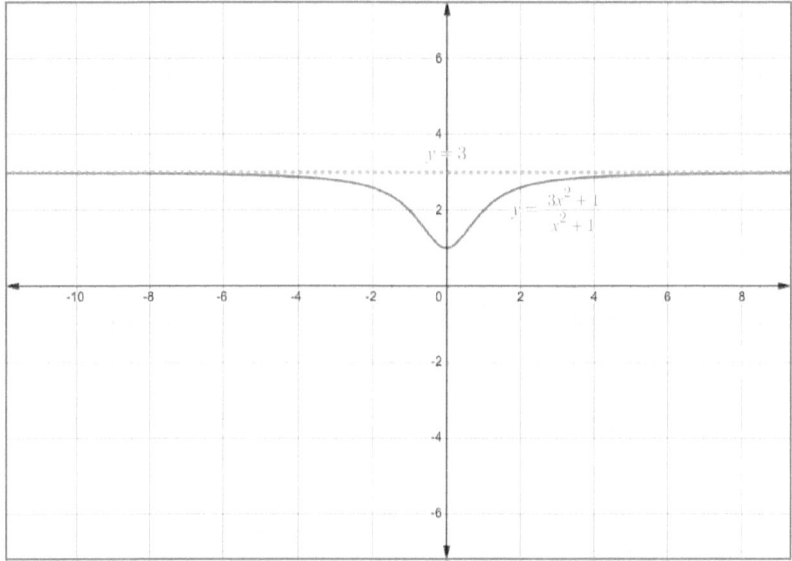

Fig. 6.6: Graph of the function $f(x) = \dfrac{3x^2+1}{(x^2+1)}$

6.7 Sharp Turn Ahead: Slant Asymptotes

Now only one case remained, that is what if the degree of the numerator is greater than the degree of the denominators (i.e. $n > m$). Again, a simple rule, you divide the numerator $p(x)$ by denominator $q(x)$, and whatever is the quotient, that line is the asymptote (Just ignore the remainder). That's it.

Exemplum 6.4: Find the slant asymptote in the graph of the function:

$$f(x) = \frac{(x^2+1)}{(x-1)}$$

Solution: The numerator has a degree 2 and the denominator has degree 1, hence, this indicates a slant asymptote (and of course there is a vertical asymptote at $x = 1$). We do the long division of polynomials.

$$\begin{array}{r} x+1 \\ x-1{\overline{\smash{\big)}\,x^2+0x+1}} \\ \underline{-x^2+x} \\ x+1 \\ \underline{-x+1} \\ 2 \end{array}$$

The quotient happens to be $x + 1$, hence the linear graph $y = x + 1$ happens to be the slant asymptote. The same is indicated in the graph of the function as shown in Fig. 6.7.

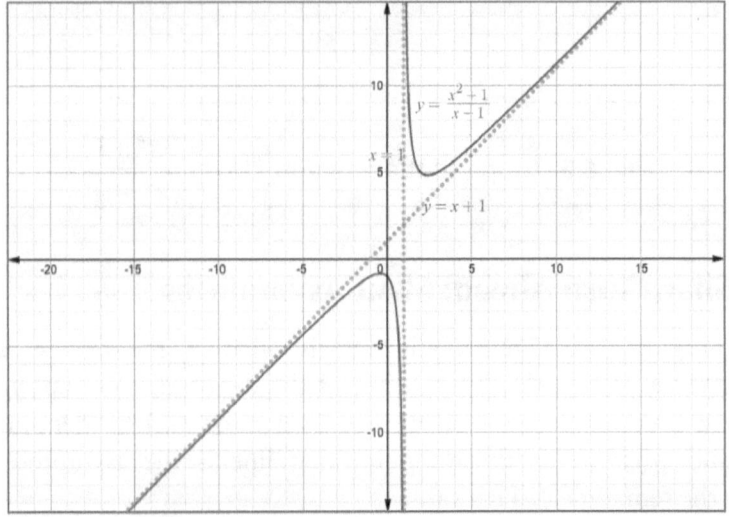

Fig. 6.7: Graph of the function $f(x) = \dfrac{(x^2+1)}{(x-1)}$

Query 6.2: I offer here a few statements which are in fact true (really, no tricks). It's just that instead of giving them as facts, I want you to prove it to yourself and be satisfied by experience. So go ahead pick any rational function that you want of any degree and have some fun.

  a. A graph can have any number of vertical asymptotes.

  b. A graph can have a maximum of only two horizontal asymptotes.

  c. A graph can have no or a maximum of 1 slant asymptote.

d. A graph can have both vertical and slant asymptote.

e. A graph cannot have both horizontal and slant asymptote

Meme 6.3: Literally never ends

6.8 Irrational Functions and the Murder of Hippasus

Hippasus of Metapontum, around 500 B.C. in ancient Greece was a member of a secret group named Pythogoreans. Of course, you can guess where the name comes from – the great mathematician Pythogoras. These were people who were obsessed with numbers – that everything in the world can be explained on the basis of numbers – weather, food, stars and that is there. However, they were particular about what kind of numbers – and that was whole numbers. Numbers that were neat and nice. In fact, they assigned genders to the numbers – the odd numbers were males and the even numbers were females. Interesting.

Pythagoras is best known for his theorem (that of course, we are going to use now and then) that helps us in calculating one side of right angled triangle if the other two are given (the famous $a^2 + b^2 = c^2$). Little did the Pythagoreans knew that this very theorem would shake their deepest beliefs. In fact, Hippasus had imagined something very terrible.

Let's say if you have a square of side 1 unit, then the question is what is the length of the diagonal?

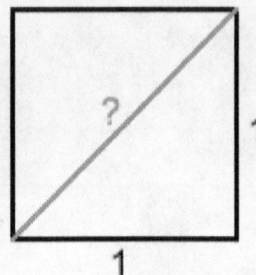

Fig. 6.8: Finding a diagonal of the square

Of course, the answer is $\sqrt{1^2 + 1^2} = \sqrt{2}$. But what is this number $\sqrt{2}$? The Pythogoreans firmly believed that this number has to be ratio of two whole numbers (for example 50 = 100/2). They were quite convinced about it. Hippasus, while sailing on a ship on coasts of Greece along with his fellow Pythogoreans had a disturbing thought. He started by thinking that indeed there are two numbers that give us the ratio of $\sqrt{2}$ and that it is a fraction in simplest terms (that means no common factors). Therefore,

$\sqrt{2} = \dfrac{\textit{some numerator}}{\textit{some denominator}}$, if we square both sides, we get

$2 = \dfrac{(\textit{some numerator}^2)}{(\textit{some denominator}^2)}$, or $2 \times (\textit{some denominator})^2 = (\textit{some numerator})^2$

But wait! That would mean (*some numerator*)² is actually an even number (since anything multiplied by 2 gives us an even number), which by the way means that (*some numerator*) is also even! (Think about it, square of odd number gives us odd numbers and that of even numbers gives us even numbers).

That would also mean that (*some denominator*) is also an even number (since square of numerator and denominator divided gives us 2, an even number). If both the numerator and denominator are even numbers, that means they both must be divisible by 2!

But wait, wait, wait! We started by saying that this ratio is in simplest terms and has no common factors and we ended up proving that they both have a factor of 2. What sort of sorcery is this? You see the problem here. We have just contradicted ourselves. Hippasus told his fellow Pythahorean friends about this disturbing discovery that perhaps there are numbers which cannot be represented in form of whole numbers that they are never-ending with no pattern in them.

"How irrational he is!" they thought and threw him right into the ocean for doubting the holy laws of mathematics. The very thought of having these kind of numbers frightened them and made them uneasy – perhaps it was "irrational" to even think about it – square roots, cube roots other roots of numbers which are not perfect, which human would even dare to think about it?

Yet, here we are – 2000 years later, acknowledging them and accepting the reality of not only irrational numbers, but also irrational functions. Formally,

An irrational function is of the type $f(x) = x^{1/n}$, where n is an integer, and not equal to zero.

Naturally, n can be either odd or even. The graphs of the same are demonstrated by the illustration given below.

Meme 6.4: The root of all rebellion is thinking

Illustration 6.3: Graphing the function $f(x) = x^{1/n}$, when n is even. Let's take $n = 2$, and then it transforms to $f(x) = x^{1/2} = \sqrt{x}$, the square root function. Of course, we cannot put negative numbers in it (at least for now, I mean, of course, we can take the help of complex numbers – but that is a whole different topic of discussion and we shall not come back to it for a while.). The rest is the usual procedure that we have been following so far, putting few numbers, getting some points (like, (0,0), (1,1), (4,2), etc.) and plotting them.

The graph has been shown in Fig. 6.9. Notice that slow growth of the graph. Does that add up? One can say that since we are not allowed to input any negative numbers, the domain has to be $[0, \infty)$ and also we do not get any negative numbers as output, hence the range happens to be $[0, \infty)$ too!

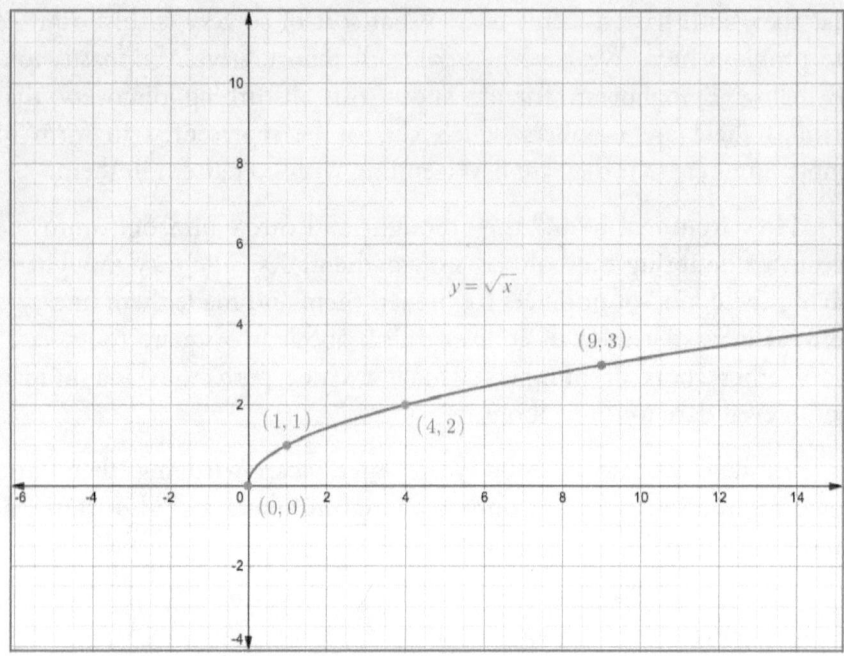

Fig. 6.9: Graph of the function $y = f(x) = \sqrt{x}$

Illustration 6.4: Graphing the function $f(x) = x^{1/n}$, when n is odd. Let's take n = 3, and then it transforms to $f(x) = x^{1/3} = \sqrt[3]{x}$, the cube root function. Of course, we can easily put negative numbers in it. Luckily, we do have cube root of negative numbers. The rest is the usual procedure that we have been following so far, putting few numbers, getting some points (like, (0,0), (−1,−1), (1,1), (−8,−2), (8,2) etc.) and plotting them.

Table 6.2: Set of values for the function $y = f(x) = \sqrt[3]{x}$

x	−27	−8	−1	0	1	8	27
$y = x^{1/3}$	−3	−2	−1	0	1	2	3

The graph has been shown in Fig. 6.10. Notice that slow growth of the graph on both sides. Does that add up? Since we are allowed to put any number as input and we get any number as output, we can say that the Domain and Range of this function as both R.

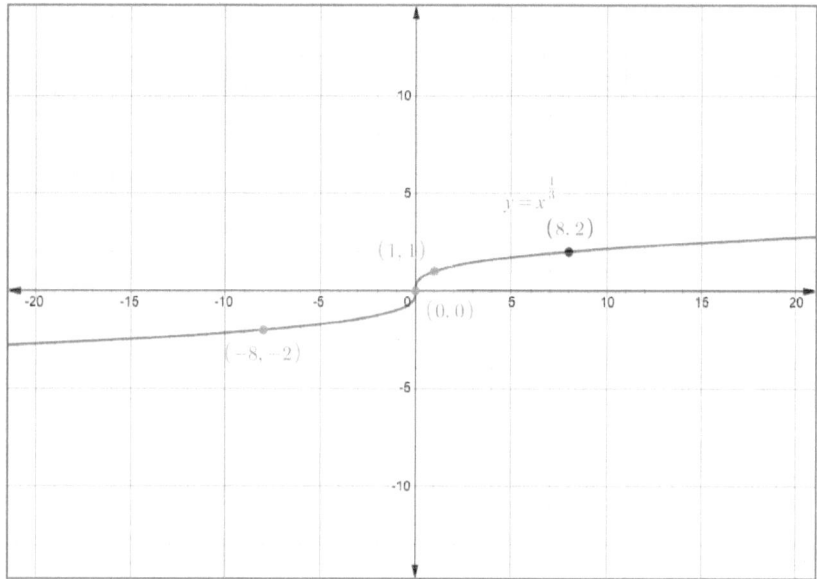

Fig. 6.10: Graph of the function $y = f(x) = \sqrt[3]{x}$

Query 6.3: Draw the graph for family of functions of even powered rational functions $y = f(x) = \sqrt[n]{x}$, where n is even. Is there any axis of symmetry?

Query 6.4: Draw the graph for family of functions of even powered rational functions $y = f(x) = \sqrt[n]{x}$, where n is odd. Is there any axis of symmetry?

6.9 To Ruin Everything You've Learned So Far

It's always fun to ruin things, isn't it? This chapter was our first introduction to asymptotes and an important part of studying Calculus. We figured that given a rational function, how do we determine horizontal, vertical and slant asymptotes. Let us try one more function as an example:

$$f(x) = \frac{(x^2 - 9)}{(x - 3)}$$

We can clearly see that the denominator will go zero at $x = 3$. Therefore, there must a vertical asymptote at $x = 3$. The degree of the numerator is greater than the denominator and that clearly indicates a slant asymptote. When we divide $x^2 - 9$ by $x - 3$, we get $x + 3$ as the quotient and hence, we should have $x + 3$ as the slant asymptote.

One last observation is that when we do put, $x = 3$, we get the form 0/0 – which is the indeterminate form! And no one knows what to do about it. Anyway, you have certain expectations about asymptotes and one point that causes some problems. Let us look at the computer generated graph of this function as shown in Fig. 6.11.

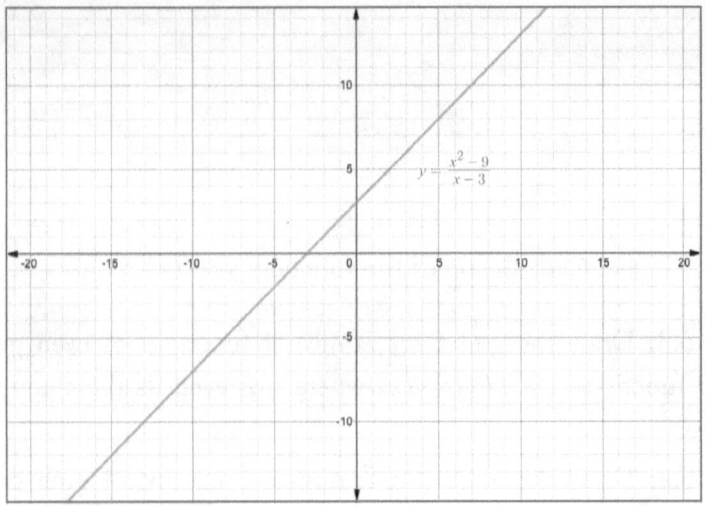

Fig. 6.11: Graph of the function $f(x) = \dfrac{(x^2 - 9)}{(x - 3)}$

What do you say after seeing the graph now? No vertical asymptote! No slant asymptote! No problem at $x = 3$. Does that mean that whatever we have studied is wrong and doesn't' work all the time? Is it?

Well, I will reserve my answer until the next book when we start Limits.

6.10 Closure

Rational and Irrational functions are found in abundance in nature. The tricky part is that they are always hidden. For example, gravitational force (which is an example of the dependence $f(x) = \dfrac{1}{x^2}$) is what keeps the world together, is a form of rational function. It is what keeps the planets, solar system and universe at the grand scale moving. Now, of course, you cannot right away guess that dependence as I have mentioned again and again. But once you start looking closely – you will see it. Electrostatic forces (because of which you feel friction) also have the dependence of $f(x) = \dfrac{1}{x^2}$. These are the forces that act at the atomic level. $f(x) = \dfrac{1}{x^2}$ is also called the inverse square law. Once you get a hold of it, you will see that it is everywhere – from infinitesimal to the infinite.

Irrational functions occur at even more intricate of places – some of which you may notice the exercises in this chapter. Naturally, irrational functions, like irrational numbers are quite uneasy to work with. It is one thing to understand them as mathematical tools and move ahead and it is another thing to actually understand them. After all, they are called 'irrational' for a reason!

Lastly, asymptotes is the most unfortunate thing that has happened to mathematicians and scientists alike. As you go ahead with calculus, you'll notice why. In fact, not only in Calculus – but various other branches of studies like finance, geology etc. As a part of closure, I advise you to play with different kind of rational/irrational functions that you can think. Read more about where they are found in nature and what things they control in your life. Trust me, you'll be surprised!

Exercises

1. Write the equation of the curve that is always 5 units away from the intersection point of the vertical and slant asymptotes of the curve $f(x) = \dfrac{x^2 + x - 6}{x - 2}$.

2. Find the domain for the following:

    a. $f(x) = \sqrt[5]{x^2 + 5x}$

    b. $f(x) = \sqrt[4]{x^2 - 8x}$

    c. $f(x) = \dfrac{x}{x + 2}$

    d. $f(x) = \dfrac{1}{\sqrt{x - 3}}$

    e. $f(x) = \sqrt{-\omega x}, (\omega > 0)$

    f. $f(x) = \sqrt{x + \sqrt{-x}}$

3. We all love sitting by the river and see the particles flow along with it. It is found that the mass of the particles that a river can take with itself depends on the sixth power of the speed of the river. Make a graph of the dependence of the mass of flowing particles with respect to the speed of the river. In how many quadrants does the graph lie?

4. An approximate form of Kepler's third law is written as:

$$T^2 = \dfrac{4\pi^2}{GM} R^3$$

Where T is the time period of the planet, R is the length of its semi-major axis and M is the mass of the planet. The value of G is 6.67×10^{-11}. Express T as a function of R. The masses of a few planets are given below:

Planet	Earth	Mars	Jupiter	Saturn
Mass (kg)	6×10^{24}	6.4×10^{23}	1.9×10^{27}	5.7×10^{26}

For each one of the planets, draw the graphs of $T = f(R)$ on the same graph paper. Do you notice any pattern or do the graphs intersect anywhere?

5. Find asymptotes, if any, in the following graphs:

   a. $\dfrac{2x}{x^2 - 5}$

   b. $\dfrac{x^2 - x - 20}{x^2 + 6x + 9}$

   c. $\dfrac{7x}{x^2 + 5}$

   d. $\dfrac{x^3 - 8}{x^2 + 2}$

   e. $\dfrac{-5x^4 - 3x^3 + x^2 - 10}{x^3 - 3x^2 + 3x - 1}$

6. If two resistors, each of resistance x ohms and y ohms are connected together in parallel, their total resistance R is given by:

$$\frac{1}{R} = \frac{1}{x} + \frac{1}{y}$$

If the total resistance always remains 2 ohms, express $y = f(x)$. Draw the graph and find its domain and range. What does the graph tell you?

7. Find a rational functions $y = f(x)$, such that it satisfies all the conditions given below:

   a. $f(3) = 0$

   b. f has a horizontal asymptote at $y = 2$

   c. f has vertical asymptotes at $x = 4$ and $x = -4$

   d. $f(0) = 1$

8. Is it possible to make a new rational function with two given rational functions? If yes, how does it affect the domain and range of the new rational function?

9. Find asymptotes of $x^2y - 9y - 4x = 0$.

10. You must have noticed standing at the railway station, that the siren of a train passing by sounds different compared to the one standing still. This change in the frequency of the sound is what we call Doppler Effect that depends on the velocity of the passing object. For a train, mathematically, it is written as:

$$f(v) = f_0 \left(\frac{s_0}{s_0 - v} \right)$$

where, f_0 is the actual frequency of the whistle of the train (let us say its 440 Hz) and s_0 is the speed of sound in air (332 m/s). Graph the function $f(v)$ What does that asymptote of the function represent physically?

7 To Map Atoms and Cosmos: Trigonometric Functions

> *"Math is the only place where truth and beauty mean the same thing"*
>
> — *Danica McKellar*

I must start this chapter with a confession. The sheer extent of the history and usage of trigonometry is impossible, just impossible to put together in one chapter. To even attempt to do so, would be complete insult to this majestic thing that is trigonometry. A separate book would have been the best way. Right from Hipparchus, Pythagoras, Aristarchus, Ptolemy, Aryabhatta, Brahmagupta, Bhaskara-I, Abu-al-wafa, Tang Dynasty, Jamshid-al-Kashi to Euler – the amount of people and generations that have worked over thousands of years to make trigonometry the way we know it today is endless. The Greeks were the original fathers of trigonometry – however their approach was so clumsy that later when the Indian method came into being, it was immediately translated into Arabic and Chinese languages for further use.

Hence, the confession being that whatever I write here will nowhere be sufficient to teach you trigonometry, and so, I restrict myself to convey to you just the 'trigonometric functions' and not trigonometry itself. And if this chapter feels like it is compressing a lot of information in one go, well that's because – it is.

7.1 The Magic of Proportions

Conceptually, trigonometry is all about proportions – the magic of shrinking and enlarging shapes and historically, trigonometry has always been about the chord of a circle (however, we are not getting into it now!). This idea of proportions leads us to the concept of similarity – that sides of a shape grow in same proportion if the corresponding angles remain the same. It is this angle staying the same that does all the trick. If we take a triangle *EFG*, with sides being *EF = 2 cm, FG = 1 cm and GE = 2 cm* and angles being $\angle E = 40°, \angle F = 70°$ and $\angle G = 70°$. Then to get a *similar triangle KLM*, we must enlarge all sides by the same amount (in our case two times), provided that the corresponding angles remain the same. And that means, if any side in *KLM* is unknown, we can just go back to our smaller triangle *EFG*, look for the corresponding side and just double it (as shown in Fig. 7.1).

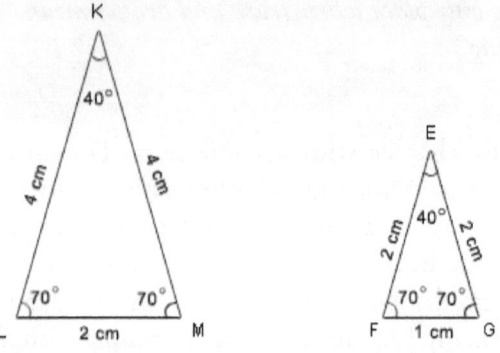

Fig. 7.1: Similar Triangles

This seemingly easy relation between angles and sides was somehow magically used by ancient Indians. Fig. 7.2 is from Jantar Mantar, Delhi – one of many observatories across India.

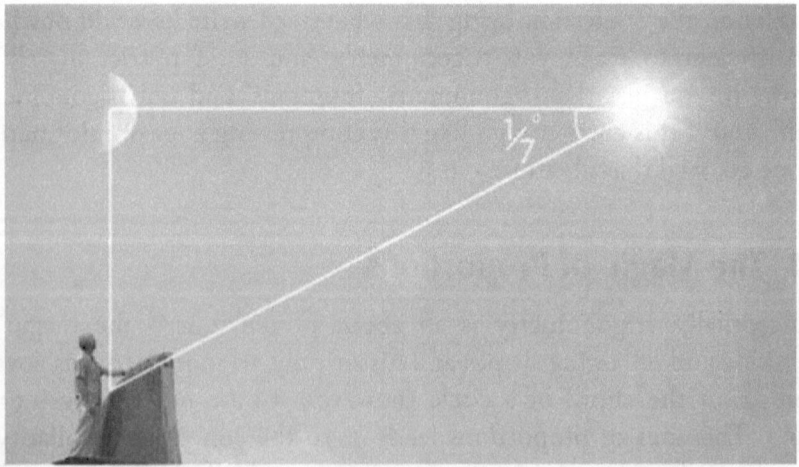

Fig. 7.2: How Indians calculated ratio of distances to sun and moon

The architectural monuments are Jantar Mantar are just giant versions of the tools of your geometry box (of course, they are a little bit more complex). The ancient Greeks and Egyptians had already calculated the distance to the moon, however, their calculations involving the distances to sun were not accurate.

Using these giant instruments, the Indians figured out that when there is a half-moon and it occurs in sky together with sun, it makes an angle of 90 degrees with us and the sun, thus forming a right-angled triangle (let's be honest here, everyone prefers a right-angled triangle because you can apply Pythagoras' theorem to it!). They measured that in such conditions, sun makes an angle of $1/7^{th}$ of a degree to an observer. What they did next will blow your mind.

They constructed a smaller version of right-angled triangle in which one of the angles was $1/7^{th}$ of a degree and found out that the adjacent side to the angle is 400 times bigger than the opposite side (I advise you to draw your own diagram of this miniature triangle). In the image shown in Fig. 7.2, the distance from the moon to us is opposite to the angle, and distance from moon to sun is the adjacent side – hence, using the idea of proportionality sun must be 400 times farther than the moon is to us. A simple magic of ratios! We don't need to go to space to know about space – in fact, magical calculations like these have actually helped us to go the space and the unknown.

The basic idea here is important – to be able to figure ratio of sides for a particular angle – which leads us to our next segment.

7.2 Aryabhatta's Trigonometric Ratios

One of the greatest Indian mathematicians, Aryabhatta (around 476–550 A.D.), who made extensive studies on astronomy and trigonometry named a few ratios in a triangle. We shall define all the possible ones here.

Let's consider a triangle ABC, right-angled at B (Fig. 7.3). The measure of $\angle C$ is called θ. If we are looking from θ, AB will be called the opposite side and BC will be called the adjacent side. AC in any case is the biggest side of the triangle called the hypotenuse (I assume that you are already familiar with these terms).

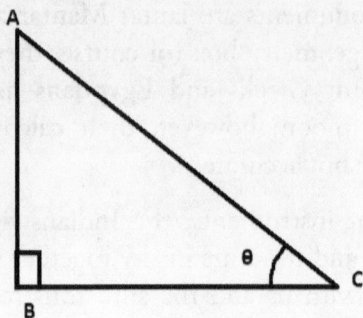

Fig. 7.3: A right-angled triangle

Now, for a given angle θ, we define the following ratios:
$$\sin\theta = \frac{opposite}{hypotenuse},\ \cos\theta = \frac{adjacent}{hypotenuse},\ \tan\theta = \frac{opposite}{adjacent},$$

And the reciprocal of each one of them as,
$$\csc\theta = \frac{hypotenuse}{opposite},\ \sec\theta = \frac{hypotenuse}{adjacent},\ \tan\theta = \frac{adjacent}{opposite}$$

Query 7.1: Which trigonometric ratio was used in determination of distance of sun with respect to the moon in the previous section?

Remember that these are simply names for all possible ratios of sides of triangle, no scary stuff.

Aryabhatta called sine as jya and cosine as kotijya. He defined the first three ratios and their reciprocals were used by Muhammad ibn Jabir al Hahaai approximately 400 years later on.

Meme 7.1: Never!

Of course, the problem is that we have to calculate these ratios for all possible angles between 0 to 360 degrees (0 to 2π radians). But then, for most angles, your calculator has to do the work. We can however, calculate manually for some of them.

Illustration 7.1: Let's try to find all trigonometric ratios for a 45-45-90 triangle. Here is the diagram of a triangle ACB (Fig. 7.4), right-angled at C. Naturally, $\angle A = \angle B = 45°$ or $\dfrac{\pi}{4}$.

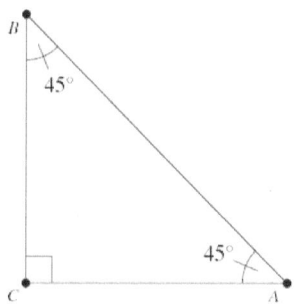

Fig. 7.4: A 45-45-90 Triangle

This also implies that $AC = CB$ (guess why?). Let $AC = CB = x$ units. In that case, simply applying Pythagoras' theorem tells us:

$AB^2 = AC^2 + CB^2$, or, $AB^2 = x^2 + x^2$, and basically, $AB = \sqrt{2}x$ units.

Therefore, $\quad \sin 45° = \dfrac{x}{\sqrt{2}x} = \dfrac{1}{\sqrt{2}}, \quad \cos 45° = \dfrac{x}{\sqrt{2}x} = \dfrac{1}{\sqrt{2}} \quad$ and $\tan 45° = \dfrac{x}{x} = 1$ and so forth for the reciprocal ratios. It's just a matter of tricks. Usually, with right-angled triangles having nice angles like 30°, 45°, 60° it is easier to find some relation between sides and angles and hence their respective trigonometric ratios. It is obviously harder with angles like 21.5°, 59° or something like that.

Whatever we have done above comes in the domain of Trigonometry – that essentially tells us about the relationship of sides with angle. The main issue of this chapter is to learn about *trigonometric functions*. A function is all about dependencies. Let's see how we transit from trigonometry to trigonometric functions.

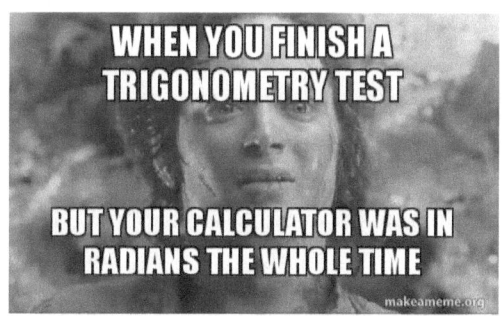

Meme 7.2: Tragedy

7.3 The Unit Circle – Birthplace of Trig-Functions

Imagine this triangle with hypotenuse of length 1 unit making θ angle (is obviously a variable) with origin and base on the x-axis. We now construct a circle with a radius equal to the length of the hypotenuse i.e. 1 unit. From cartesian system, it's quite obvious that the side adjacent to θ (i.e. base of the triangle) will be the measure of *x-coordinate* and the side opposite to θ (i.e. the perpendicular) will be the measure of the *y-coordinate* of the point P, where the triangle touches the circle. We know that,

$$sin\ \theta = \frac{opposite}{hypotenuse},\ cos\ \theta = \frac{adjacent}{hypotenuse}$$

And in our case, the hypotenuse is just 1, hence we can write:

sin θ = *opposite (or the perpendicular)*, *cos* θ = *adjacent (or the base)*, as shown in Fig. 7.5

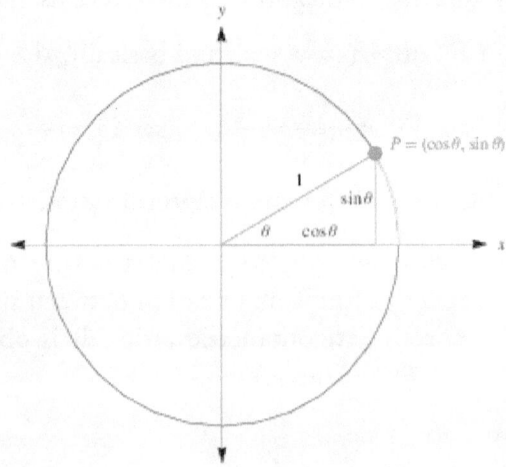

Fig. 7.5: Unit Circle

This now is a new way of looking at trigonometric functions. The point $P(x,y)$ is simply now $P(\cos\theta, \sin\theta)$ i.e. in a unit circle sine of an angle is simply the length of perpendicular of the triangle enclosed in unit circle and cosine of an angle is simply the length of base. Their values depend on the value of θ, and thus they have now truly transformed into functions of θ, $f(\theta)$. We can calculate the length of the perpendicular and base for several values of and write them down on the unit circle.

The values for different angles and their corresponding coordinates (sine and cosine values) are shown in fig. 7.6.

Remember, you do not need to mug up the whole unit circle. Just keep coming back to it as you solve problems and questions and the values will automatically come to you. Rote learning and mugging have no place in actual mathematics.

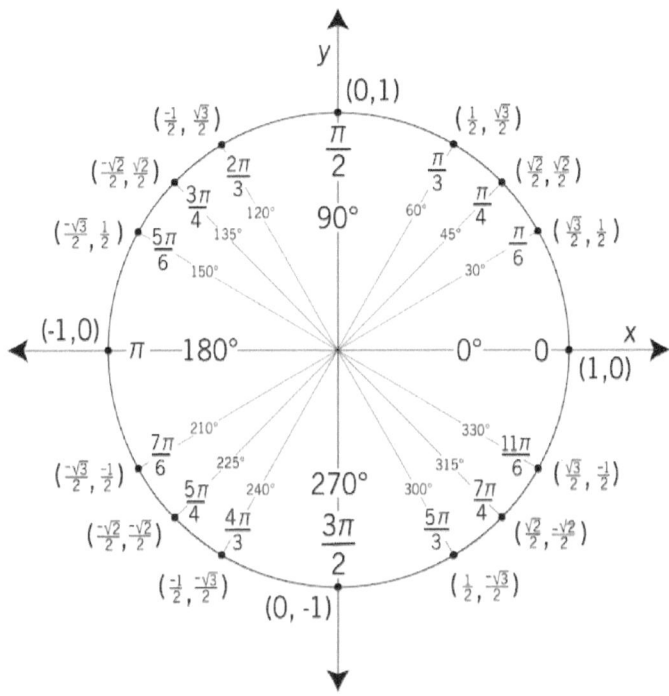

Fig. 7.6: Unit circle

Since the sine and cosine of an angle can be directly seen from the unit circle, therefore, the tangent of an angle can be indirectly obtained – simply by dividing the sine and cosine values, so is the case with other trigonometric ratios.

A new thing to notice here: Suppose we ask what is the value of sine function at 390°? To reach 390°, we must go a complete round the circle (i.e. 360°) and then move 30° clockwise; and so the sine function will have same value at 30° or 390°, because they are literally the same triangles and same coordinates. This is true for all other trigonometric functions as well. That is, they all repeat their values after a certain amount of angle (not necessarily 360° all the time). We call them as *periodic functions*. Formally,

A periodic function is a function whose value repeats itself after regular intervals (also called periods).

And suddenly, trigonometric functions can just do more than what they were already doing. If trigonometry was all about linking angles to the sides and using them for indirect measurement, trigonometric functions can now be used to define relations/dependencies in nature that are periodic in nature. Look around you, there are numerous things which are periodic in nature-oscillation of the pendulum, motion of planets, day and night, seasons, tidal waves, and even your own heartbeat! Can all of these periodic events be described by trigonometric functions? In fact, the input need not be an angle itself, it can be anything else. Think about it.

7.4 The Sine Function

In the unit circle, sin θ basically a function that tells you the length of the perpendicular of the right angled-triangle of measure θ or just the *y-coordinate*. Go back to the unit circle and see how the values of the *y-coordinate* vary with the changing θ. By defining $y = f(x) = \sin x$ (where x being the input as angle, and y being the output as the length of perpendicular), we can say the following.

In the first quadrant, at $x = 0°$, $y = 0$ (that is the when angle is zero, the length of perpendicular is zero, makes sense!), and then it increases in the first quadrant until it comes to its full length i.e. at $x = 90°$, $y = 1-$ where it is equal to the hypotenuse. This is the biggest that *y-coordinate* can be. As soon as it enters the second quadrant, the *y-coordinate* starts to decrease until it becomes zero again at $x = 180°$, $y = 0$. Thereafter, it enters the third quadrant and starts increasing but in the negative direction, becoming more and more negative until $x = 270°$, $y = -1$. This is the smallest sine function can be. Lastly, the moment it enters fourth quadrant, it is still negative but becomes smaller in size and closer to zero until $x = 360°$, $y = 0$. Back to where we started. Hereafter, sine function will repeat this entire cycle again and again every 360° or 2π radians.

Therefore, the period of sine graph is 360° or 2π radians. Its values are positive in 1st and 2nd quadrant but negative in 3rd and 4th quadrant (as expected of *y-coordinate*). The maximum value of this function is 1 and the minimum value is −1 and the function "oscillates" between these max. and min. values. There seems to be no angle that creates any problem to this function. Hence, the domain is R and the range is [−1,1].

We can use the unit circle to get values for 0 to 360° as shown in Table 7.1 and the graph as shown in Fig. 7.7.

Table 7.1: Values of sin function

x (rads)	0	$\pi/4$	$\pi/2$	$3\pi/4$	π	$5\pi/4$	$3\pi/2$	$7\pi/4$	2π
$y = \sin x$	0	$1/\sqrt{2}$	1	$1/\sqrt{2}$	0	$-1/\sqrt{2}$	−1	$-1/\sqrt{2}$	0

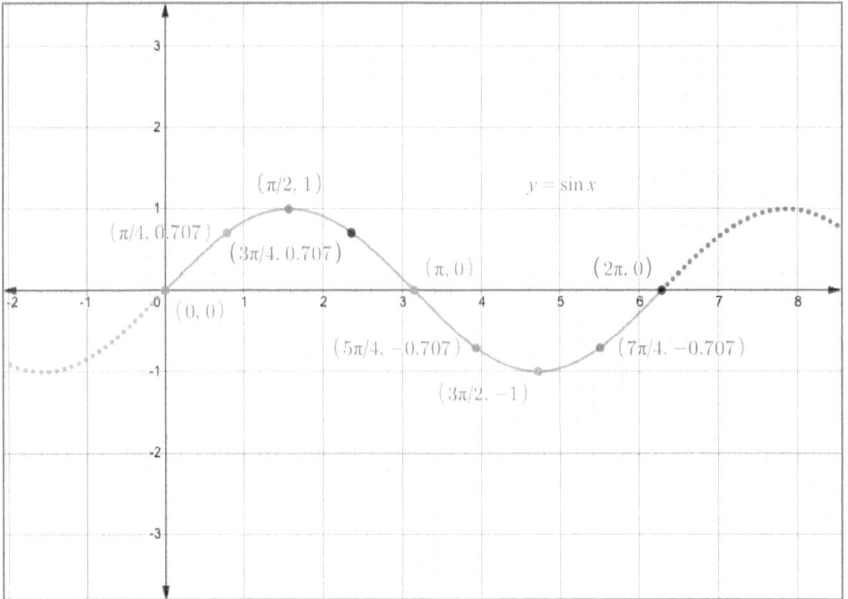

Fig. 7.7: Graph of y = sin x

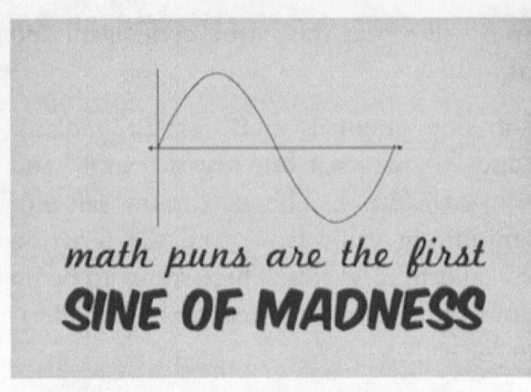

Meme 7.3: Getting in'sine'ly mad

A nice wave-shaped curve. Much different from what we have seen previously. The sine function is restricted to move in between −1 and +1 only. It just can't go beyond that. We say that the *amplitude* of the function is 1 (as that's how far it goes on both sides).

Query 7.2: Does it make sense to say that *sin (2π + θ) = sin θ*?

7.5 The Cosine Function

In the unit circle, *cos θ* basically tells you the length of the base of the right angled-triangle of measure θ or just the *x-coordinate*. Go back to the unit circle and see how the values of the *x-coordinate* vary with the changing θ. By defining $y = f(x) = \cos x$, where x being the input as angle, and y being the output as the length of base (**CAUTION**: Do not get confused here with the notation. We have used x as input which is an angle and y as the output of the value of x-coordinate or the length of the base), we have:

In the first quadrant, at $x = 0°$, $y = 1$ (that is the when angle is zero, length of base is maximum), and then it decreases in the first quadrant until it reduces to zero i.e. at $x = 90°$, $y = 0$– but remember this is not its minimum value. As soon as it enters the second quadrant and the *base* starts to increase but in a negative direction until becomes the minimum that it can be at $x = 180°$, $y = -1$. Thereafter, it enters the third quadrant and starts increasing in value, becoming more and more closer to 0 until $x = 270°$, $y = 0$. Lastly, the moment it enters fourth quadrant, it again becomes positive and gets closer to the maximum that it can be until $x = 360°$, $y = 1$. Back to where we started. Hereafter, cosine function will repeat this entire cycle again and again every 360° or 2π radians.

Therefore, the period of sine graph is 360° or 2π radians. Its values are positive in 1st and 4th quadrant but negative in 2nd and 3rd quadrant (as expected of *x-coordinate*). The maximum value of this function is 1 and the minimum value is −1 and the function "oscillates" between these max. and min. values. There seems to be no angle that creates any problem to this function. Hence, the domain is R and the range is $[-1,1]$.

We can use the unit circle to get values for 0 to 360° as shown in Table 7.2 and the graph as shown in Fig. 7.8

Table 7.2: Values of cosine function

x (rads)	0	$\pi/4$	$\pi/2$	$3\pi/4$	π	$5\pi/4$	$3\pi/2$	$7\pi/4$	2π
$y = \cos x$	1	$1/\sqrt{2}$	0	$-1/\sqrt{2}$	−1	$-1/\sqrt{2}$	0	$1/\sqrt{2}$	1

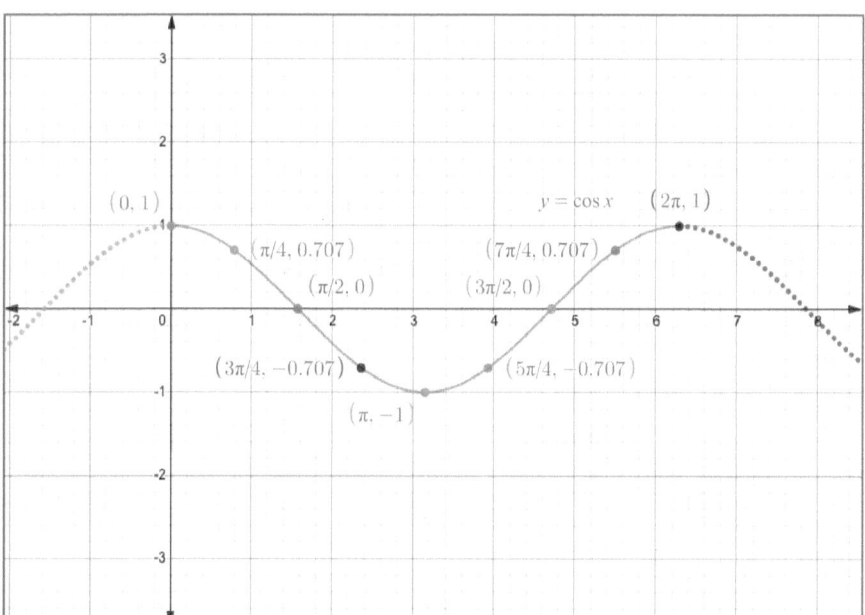

Fig. 7.8: Graph of $y = \cos x$

Meme 7.4: Cosine and Chill

Again a nice wave-shaped curve. Pretty much looks the same as the sine function but seems a bit shifted. The sine function is restricted to move in between −1 and +1 only. It just can't go beyond that. Again, the amplitude of the function is just 1.

Query 7.3: Does it make sense to say that $cos(2\pi + \theta) = \cos \theta$?

7.6 The Tangent Function

Let's make things a little bit spicier now and talk about the tangent function (does it have to do anything with tangent to a circle?). There are few things in a movie that we don't notice but in the climax they are the ones that break the entire suspense – in the movie called Calculus, tangent function is that thing. But anyway, we know that:

$$tan\theta = \frac{sin\theta}{cos\theta}$$

Interesting, there's a cos θ in the denominator. And what's the game with denominators? Everytime the denominator goes to zero, there is a vertical asymptote. Therefore, tan function has a problem every time θ is an odd multiple of 90° or $\frac{\pi}{2}$ (as cos θ becomes zero). About the signs of tangent function in different quadrants, it has been shown in Table 7.3.

Table 7.3: Signs of tangent function

Quadrant	sin θ	cos θ	tan θ
I	+	+	+
II	+	−	−
III	−	−	+
IV	−	+	−

What can we say about the values of tan function in different quadrants? Does it also stay enclosed between −1 to 1 like sine and cosine do? We use the values given in unit circle and try to understand how tangent values change between different quadrant as discussed in Table 7.4.

Table 7.4: Values of tangent function

Quadrant I	The sin values start from zero and start going towards 1 while cosine values start from 1 and start going to zero. By the definition of *tan θ*, the numerator keeps getting bigger and the denominator keeps getting smaller; thus, turning the over-all ratio gradually bigger and bigger. Therefore, *tan θ* starts from 0 and keeps increasing forever until at $\theta = \frac{\pi}{2}$, *cos θ* simply becomes zero and the whole ratio becomes undefined (or simply say, there is an asymptote).
Quadrant II	Now, The sin values start from 1 and start decreasing towards 0 (but is still positive) while cosine values start from 0 and start going to −1. By the definition of tan θ, the numerator keeps getting smaller and the denominator keeps increasing in the negative direction; thus, turning the over-all ratio a negative number. Therefore, *tan θ* starts from a very small number and keeps increasing until at $\theta = \pi$, simply becomes zero and the whole ratio becomes zero.
Quadrant III	Next, the sin values start from zero and start going towards −1 while cosine values start from −1 and start going to zero. And since both the numerator and denominator are negative, the overall ratio is positive. So, we have scenario exactly as seen in Quadrant I until at $\theta = \frac{3\pi}{2}$, *cos θ* simply becomes zero and the whole ratio becomes undefined (or simply say, there is an asymptote).
Quadrant IV	Finally, I leave this up to you to analyze why the situation here will be the same as the situation in Quadrant II.

On the basis of discussion in Table 7.4, we draw the graph of tan function as shown in Fig. 7.9.

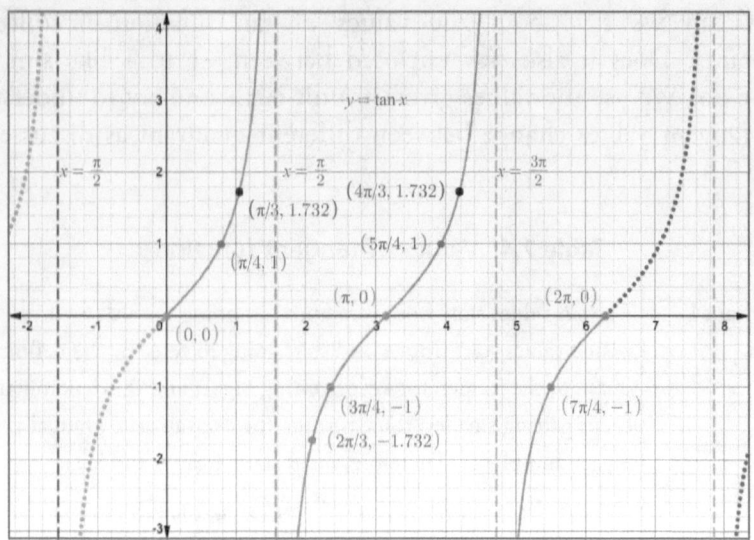

Fig. 7.9: Graph of of $y = \tan\theta$

What a weird graph! But anyway, we have to deal with it now. It seems to us that graph repeats itself not by a period 2π of like cosine and sine but rather with a period of π. Talking about the domain, any values except for the odd multiples of $\dfrac{\pi}{2}$ don't really create any problems for us. Therefore, Domain is $R - \{(2n+1)\dfrac{\pi}{2}\}$ and the Range is set of all real numbers R.

Query 7.4: Does it make sense to say that $\tan(\pi + \theta) = \tan\theta$? What is the amplitude of this function?

7.7 Bunch of Formulas

Well, of course, the dreaded section – FORMULAS! Everyone hates them. But here's the thing – they are merely here to make our life simple. There are tons of formulas in trigonometry. Remember when I said in the starting that I need to write a separate book on trigonometry – it was because of the sheer amount of the formulas involved. Here, we shall only discuss the ones that are useful to us, avoiding the proof in this chapter. What's important to understand is to have an intuition about which ones to use and when. That of course, depends on the problem at hand – but nevertheless – I am still not asking you to mug them up. You just keep playing with them until you get the hang of it and that will be enough.

7.7.1 Basic Trigonometric Formulas

Here is the first set – what we call them as the basic trigonometric formulas. There are three of them, written as:

$$sin^2 x + cos^2 x = 1$$

$$1 + tan^2 x = sec^2 x$$

$$1 + cot^2 x = cosec^2 x$$

Now, if you look closely, they are nothing but Pythagoras' theorem all over again in a fancy manner. Let us examine them in illustration 7.2

Illustration 7.2: Look back at the right-angled triangle shown in Fig. 7.2 again. If we apply Pythagoras' theorem on that triangle again, we get:

$$AB^2 + BC^2 = AC^2$$

If we divide each side by AC^2, we get:

$$\left(\frac{AB}{AC}\right)^2 + \left(\frac{BC}{AC}\right)^2 = 1$$

But, $\frac{AB}{AC}$ is simply $sin\ \theta$ and $\frac{BC}{AC}$ is simply $cos\ \theta$, therefore, we can write again:

$$sin^2 \theta + cos^2 \theta = 1$$

And that's all! See, as I told you – it is nothing but Pythagoras' theorem in a fancy form.

For the next two formulas, it should not be too tough to work that we divide $AB^2 + BC^2 = AC^2$ by BC^2 to get $1 + tan^2 \theta = sec^2 \theta$ and by AB^2 to get $1 + cot^2 \theta = cosec^2 \theta$. Simple!

The goal is not to show you the proof here, but rather to tell you that if you analyze carefully enough – these formulas will slowly start making intuitive sense to you.

7.7.2. Sum and Difference Formulas

These set of formulas are particularly useful when you do not the sine and cosine values of a particular angle but you do happen to know the sine and cosine values of two other angles whose sum/difference lead you to your required angle. For example, let's say we need to find $sin\ 75°$. Now, we do know that 75° is just equal to the addition of 45° and 30° and we also happen to know the cosine and sine values of 45° and 30°. We can thus make use of this set of formula given below:

$$sin(A+B) = sinA.cosB + cosA.sinB$$

$$sin(A-B) = sinA.cosB - cosA.sinB$$

$$cos(A+B) = cosA.cosB - sinA.sinB$$

$$cos(A-B) = cosA.cosB + sinA.sinB$$

Exemplum 7.1: Find the value of

Solution: We know that 15° = 45° − 30°, and we happen to know what $sin\ 45°$, $cos\ 45°$, $sin\ 30°$ and $cos\ 30°$ are. Thus, we take $A = 45°$ and $B = 30°$ and use the formula:

$$sin(A-B) = sinA.cosB - cosA.sinB$$

$sin\ 15° = sin\ (45° - 30°) = sin\ 45°.cos\ 30° - cos\ 45°.sin\ 30°$, or

$$sin\ 15° = \frac{1}{\sqrt{2}} \cdot \frac{\sqrt{3}}{2} - \frac{1}{\sqrt{2}} \cdot \frac{1}{2}$$

$$\therefore sin\ 15° = \frac{\sqrt{3}-1}{2\sqrt{2}}$$

7.7.3 Double Angle Formulas

As the name suggests, these help you to convert between trigonometric ratios of an angle and twice its value. Given below:

$$sin\ 2x = 2.sinx.cosx$$

$$cos\ 2x = cos^2x - sin^2x = 2cos^2x - 1 = 1 - 2sin^2x$$

But they don't do only this much! Notice that formulas $\cos 2x = 1 - 2\sin^2 x$ and $\cos 2x = 2\cos^2 x - 1$ they help you bring down a second degree term to a first degree form. This is of immense use when the problems become complicated and cannot be solved in the second degree – we then compromise on the value of the angle but are able to bring it down to the first degree.

Query 7.5: Can you find the formulas for *tan* $(A + B)$ and *tan* $2x$?

7.7.4 Product to Sum formula

Again, as the name suggests, these set of formulas, help you convert multiplication involving trigonometric formulas to addition. They are given below:

$$\sin A.\cos B = \frac{1}{2}[\sin(A+B) + \sin(A-B)]$$

$$\cos A.\sin B = \frac{1}{2}[\sin(A+B) - \sin(A-B)]$$

$$\sin A.\sin B = \frac{1}{2}[\cos(A-B) - \cos(A+B)]$$

$$\cos A.\cos B = \frac{1}{2}[\cos(A-B) + \cos(A+B)]$$

Illustration 7.3: Let's go back to the unit circle as shown in Fig. 7.5. Recall what $\sin \theta$ tells you here?

It tells us the length of the perpendicular of the unit circle as the angle θ changes. The length of the perpendicular "oscillates" between $+1$ to -1 with the change in angle. If the radius of the circle was not 1 but rather some other length, say A, the perpendicular y would oscillate between $+A$ and $-A$ and it would be represented by $y = A\sin \theta$ instead of just $\sin \theta$ (makes sense?).

Now, let's take it a step further. How fast does θ itself change? Don't know, maybe with some speed ω radians per second (don't worry, just a fancy greek symbol). In that case, θ can be written as θ, $\omega t = t$ is the time elapsed.

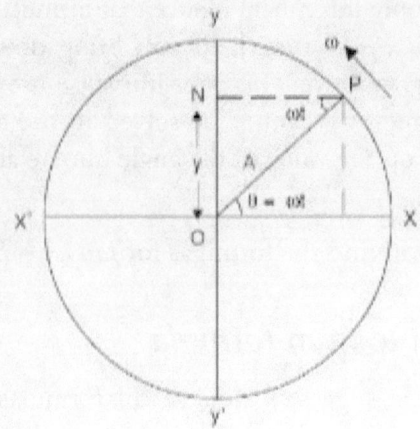

Fig. 7.10: How sin function leads to the concept of oscillation

So, in a way, $y = A\sin\theta$ can simply be written as $y = A\sin\omega t$. Focus on where this equation leads you to. It need not be restricted to length of perpendicular alone – in fact, it can tell about anything "oscillating" with time from one value +A to another value –A, kind of mathematical form of the to and fro motion.

Looking at the graph of sin function, we can say that an oscillating object following the function $y = A\sin\omega t$ would start from its "zero" position at $t = 0$, then slowly go up to the value A, come back to zero position, continue moving back to –A position and finally come back to where it started from.

Can you tell what difference it would make in the nature of oscillation if we were to write $y = A\cos\omega t$ instead of $y = A\sin\omega t$?

Exemplum 7.2: Prove that $sin2x = \dfrac{(2tanx)}{(1+tan^2 x)}$

Solution: The L.H.S. here already is simplified so the only logical conclusion would be to start from the R.H.S. and prove it equal to the L.H.S. Hence,

R.H.S. $= \dfrac{2\tan x}{1+\tan^2 x}$

One instinct upon immediately seeing tan x is to convert it to sin x and cos x. Well, we follow that instinct here (won't work every time). We get,

R.H.S. $= \dfrac{2\dfrac{\sin x}{\cos x}}{1+(\dfrac{\sin^2 x}{\cos^2 x})}$

$= \dfrac{2\dfrac{\sin x}{\cos x}}{\dfrac{\sin^2 x + \cos^2 x}{\cos^2 x}}$ [Taking L.C.M. in the denominator]

$= \dfrac{2\dfrac{\sin x}{\cos x}}{\dfrac{1}{\cos^2 x}}$ [From the basic trig. formulas]

$= \dfrac{2\sin x}{\cos x}.\cos^2 x$

$= 2\sin x.\cos x$ [Double angle formula]

$= \sin 2x =$ L.H.S.

Hence, proved.

7.8 Closure

Trigonometry, rightly said so, is the science of indirect measurements. Perhaps there is no other branch of mathematics that has expanded over such vast period in history and countries – touching all ancient civilizations. Trigonometry is directly used in astronomy, physics, archaeology, navigation, flight engineering, biology, marine engineering, constructions and possibly everything you can think of. Trigonometric functions are used to model oscillations, wave behavior and many such things.

Trigonometric functions form an important part of the study of Calculus and it is extremely important for any to be good at it in order to be good at Calculus. They come with unique properties of their own and are very different from the other transcendental/algebraic functions we know. In fact, I would go to the extent of saying that man would still be a cave dwelling creature if not for trigonometry.

You have been introduced to certain properties, graphs and formulas in trig. However, there is a lot more to it. I would seriously urge you to study trigonometry from as many sources as possible. No matter how much you do it, I assure you – it will always be less.

Exercises

1. If $3x = 4$ and $x = \sin\theta$, is it possible to solve this equation? If yes, then how? If no, then why not?

2. What are the values of $\sin(-\theta)$, $\cos(-\theta)$ and $\tan(-\theta)$?

3. If a ball is thrown at a speed of u m/s with an angle θ to the horizontal, then range R of the ball is given by the formula:
$$R = \frac{2u\sin\theta.\cos\theta}{g}$$
Choose any value of u and let the value of g be 10. Draw the graph $R = f(\theta)$. What can you say about the range of ball thrown at an angle θ and a ball thrown at an angle $45° + \theta$?

4. The phases of the moon as seen from the Earth is modelled by the dependence:
$$F = \frac{1}{2}(1 - \cos\theta)$$
where, F is the fraction of the moon that is visible on the Earth and θ is the angle between the sun, moon and the earth. Draw a plot of $F = f(\theta)$. Find the value of θ when there is:

   a. No moon

   b. Quarter Moon

   c. Half Moon

   d. Full Moon

5. Find the value of a) $\sin 8°.\cos 22° + \cos 8°.\sin 22°$ b) $\cos 22.5°$.

6. Find the domain of the following:

   a. $\sqrt{2\sin 2x} + \sqrt{25 - x^2}$

   b. $\dfrac{1}{\sin x}$

7. Prove the following:

   a. $\cos^3 2x + 3\cos 2x = 4(\cos^6 x - \sin^6 x)$

   b. $\dfrac{2(\tan x - \cot x)}{\tan^2 x - \cot^2 x} = \sin 2x$

   c. $\dfrac{\sin 3x}{\sin x . \cos x} = 4\cos x - \sec x$

8. Make a research/report on the Great Trigonometrical Survey of India – the story of how India was measured. Explain the role of trigonometry in the Survey.

8 Life Is All About Growth: Exponential and Logarithmic Functions

"Exponentials can't go on forever, because they will gobble up everything."

– *Carl Sagan*

We rarely expect mathematics to creep in our mythological stories and folktales. I remember this particular story that my grandpa used to tell me. For starters, it only seemed to talk about the moral lesson at the end of it. But growing up, I realized how it was more of mathematics. Now of course, I claim by no means that it actually happened in reality but it's intriguing to think that this story simply exists. It's about a dish called Payasam (also called Kheer in north India – a dish that is made up of rice and milk), chess and surprisingly exponential functions!

8.1 Payasam, Chess and Exponential Functions

Legend has it that in an ancient town of Kerala, India, God himself appeared in the form of Sage to a wealthy and proud king. The Sage challenged the king for a game of chess. The king, being an extremely exuberant fellow and a chess enthusiast agreed to play with the condition being that whoever loses the game must give the other what he asks for – and there was nothing in the world that the king would not have been able to offer!

The king, ultimately, lost the game. "Ask what you may! I will offer you whatever riches you may ask!", said the king trying to uphold his lost pride. The Sage made a rather modest request – "I just need some rice to make Payasam!". "That's it?", the king said, "that's all you ask amongst all the money and gold that I can offer you?". "Yes", the sage said, "however I have one request". The king, almost shocked and unable to control his laughter said, "Go ahead". The Sage replied, "I don't want all the rice in one ago, I want you to give me the rice on this chess – 1 grain today on the first square, 2 grains tomorrow on the second square, 4 grains the day after on the third square and so on."

Fig. 8.1: The chessboard and the grains

It was rather a poor and an odd request. However, the king agreed. He was not realizing what he got into. There was obviously a pattern. The Sage demanded grains in a pattern of 2^x, where x is the number of square of on the chessboard. Within a month, his stock keeper came running to him that the entire rice stocks of the royal stock has finished. Let's see the mathematics of it – for about 32 days, the number of rice grains would be: $2^0 + 2^1 + 2^3 + 2^4 + ... 2^{31}$ and that happens to be 4, 294, 967, 295! If you take the weight of each grain of rice to be about 65 grams, then that means in about a month, the total weight of rice given to the sage would be about 279,000 kilograms! If you continue the pattern to all the 64 squares of the chessboard, the total number of grains would be: $2^0 + 2^1 + 2^3 + 2^4 + ... 2^{63}$ = 18, 446, 744, 073, 709, 551, 615 grains of rice. And if you don't know how much that weighs – then let me tell you that it is roughly 1500 times the consumption of the entire world! You think the king would have ever been able to give away that?

The Sage had trapped him with what we call here as Exponential functions. He started with some value a grain of rice (1 grain in this case) and continued multiplying a number *b* (2 in our story) and the total went beyond what one could expect. To define it formally,

An exponential function is of the type $y = f(x) = ab^x$, where $b > 0$.

All kinds of growth and decay in nature are modelled via this function. Remind yourself what you read about the number e in Chapter 3.

Query 8.1 Explain why in the exponential function $y = f(x) = ab^x$, *b* is not allowed to be negative?

8.2 Exponential Growth Functions

In the exponential function $y = f(x) = ab^x$, if b > 1, then it is called an exponential growth function. The value of *a* is what we call as an initial value and *b* is called the growth factor. If you plug, $x = 0$, you get $y = a$ or simply the point *(0, a)*, so no matter the graph of an exponential function, it is always going to pass through this point.

To get a feel, let's start by plugging $a = 1$, and let's say $b = 2$ (I was so tempted to take $b = e$ here but let's keep the values simple). Let's the graph of this function $y = f(x) = 1.2^x = 2^x$.

Illustration 8.1: To graph the exponential that our sage used, $y = f(x) = 2^x$. We start as usual by representing our function in numerical form as shown in Table 8.1.

Table 8.1: Values for the function $f(x) = 2^x$

x	−4	−3	−2	−1	0	1	2	3	4
$y = 2^x$	0.0625	0.125	0.25	0.5	1	2	4	8	16

Few things to notice: growth is clearly shown here, as *x* increases, *y* increases rapidly too. Secondly, there is no as much restriction to what value of *x* we input. A negative value of *x* gives a small value and as *x* becomes more and more negative, the values become really really small – almost trying to touch zero, but never really touching it.

Next thing, if we try to graph this function back to the cartesian plane, we get the graph as shown in Fig. 8.2.

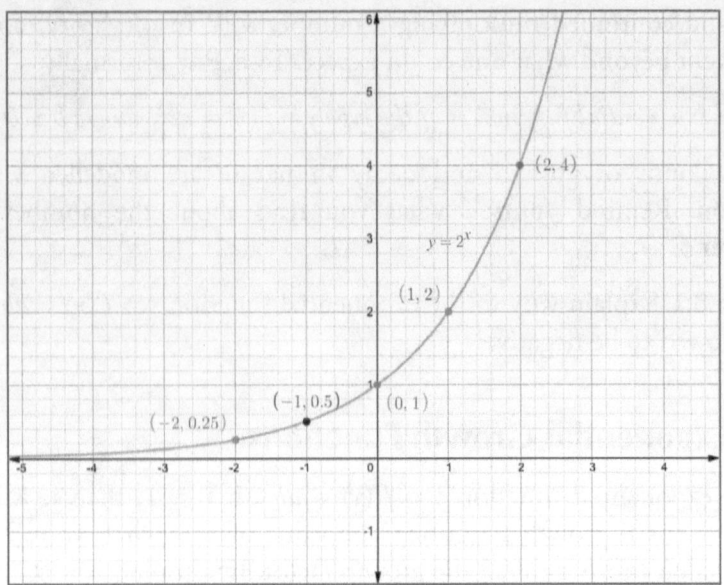

Fig. 8.2: Graph of the function f(x) = 2^x

A nice, neat exponential curve – which goes on increasing. We can do the same with the other values of *b* and get a *family of exponential growth functions*. The graph for a few exponential curves is shown in Fig. 8.3

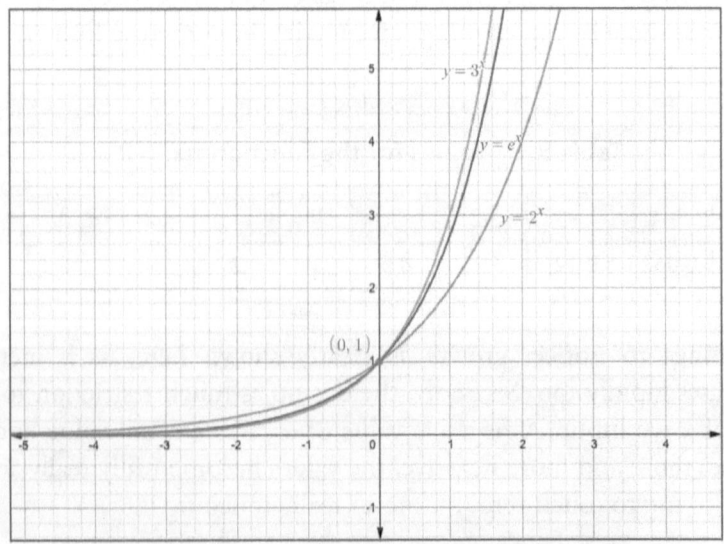

Fig. 8.3: Graphs of the functions f(x) = 2^x, e^x, 3^x

By now, it should be obvious that the range of exponential growth functions is (0,∞) and the domain is R. Notice that the x-axis is the horizontal asymptote.

8.3 Exponential Decay Functions

In the exponential function $y = f(x) = ab^x$, if $0 < b < 1$ (basically a decimal value), then it is called an exponential growth function. The value of a is what we call as an initial value and b is called the decay factor.

Again, let's start by plugging $a = 1$, and let's say $b = 1/2$. Let's see the graph of this function $y = f(x) = 1 \cdot \left(\frac{1}{2}\right)^x = \left(\frac{1}{2}\right)^x$. We again, use an illustration to draw the graph of this function.

Illustration 8.2: To graph the exponential that our sage used, $y = f(x) = \left(\frac{1}{2}\right)^x$. We start as usual by representing our function in numerical form:

Table 8.2: Values for the function $f(x) = (1/2)^x$

x	−4	−3	−2	−1	0	1	2	3	4
$y = \left(\frac{1}{2}\right)^x$	16	8	4	2	1	0.5	0.25	0.125	0.0625

Few things to notice, growth is clearly shown here, as x increases, y decreases rapidly. Secondly, there is no as much restriction to what value of x we input. A positive value of x gives a small value and as x becomes more and more positive, the values become really really small – almost trying to touch zero, but never really touching it. It is almost like the values in the graph $y = 2^x$ of have been reversed.

Next thing, if we try to graph this function back to the cartesian plane, we get the graph as shown in Fig. 8.4.

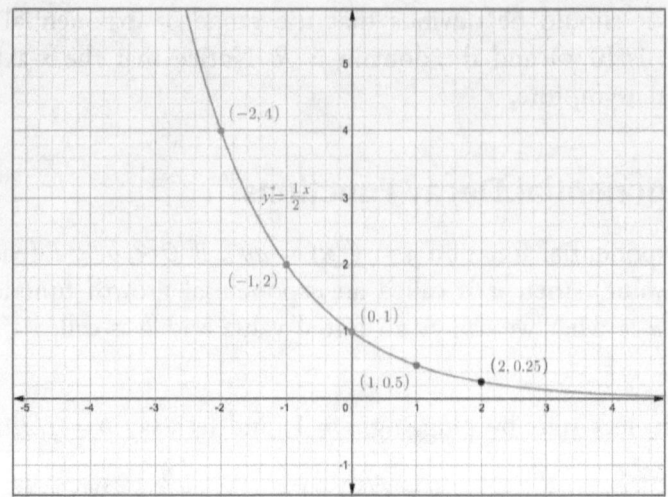

Fig. 8.4: Graph of the function $f(x) = (1/2)^x$

Notice and that the range and domain remain unchanged here in comparison to the exponential growth function.

Query 8.2 Draw the graphs of family of exponential decay functions.

Exemplum 8.1: Find the domain and range of the function $f(x) = 4.2^x + 3$.

Solution: The range of the basic exponential function 2^x is $f(x) > 0$. Multiplying 4 doesn't really change it. However, adding a 3 means that the lowest value is now $0 + 3 = 3$, hence the range is $(3, \infty)$. Regarding the input, there seems to be no value of x that causes any problem, hence the Domain is still R.

8.4 Natural Base Exponential Function

Remembering our discussion on the number e in Chapter 3, it should not come as a surprise that e^x is obviously special. But the story doesn't end here – you should be prepared to have your mind blown when you go to Vol. 2 of this book. Till then, just hold on to the suspense. A lot phenomenons that grow rapidly like growth of a bacteria, decay of a radioactive sample can all be modelled by this function.

Query 8.3 Are the numbers π^e, e^e, and e^π, algebraic or transcendental?

Exemplum 8.2: Solve the equation $2^{2x} + 32 = 12.2^x$

Solution: At first glance, this looks like a weird question about exponents. But things get easier once you notice 2^{2x} is simply the square of 2^x. So it basically means that if you call 2^x as some variable y, then 2^{2x} is automatically y^2. Then the equation becomes:

$y^2 + 32 = 12y$ which is a simple quadratic equation that can be solved by factorization!

$$\Rightarrow y^2 - 12y + 32 = 0$$
$$\Rightarrow y^2 - 4y - 8y + 32 = 0$$
$$\Rightarrow y(y-4) - 8(y-4) = 0$$
$$\Rightarrow (y-4)(y-8) = 0$$
$$\Rightarrow y = 4 \text{ or } y = 8$$

But that's not the end of the problem. We merely wrote 2^x as y in order to make the problem look more comfortable. Therefore,

$$\Rightarrow 2^x = 4 \text{ or } 2^x = 8$$
$$\Rightarrow 2^x = 2^2 \text{ or } 2^x = 2^3$$

Therefore, $x = 2$ or $x = 3$.

8.5 Watchmaker, Laird and Logarithms

What happens when you multiply 1000 × 10000? Of course, you can go about the long way – doing the long multiplication. But there seems to be an obvious trick to this. 1000 is just 10^3 and 10000 is just 10^4. The answer which 10000000, can simply be written as 10^7, which is the addition of exponents 3 and 4. Obviously, it is easy with the powers of 10, but can't be done with problems like 23535 × 3627743. So, the question is – can you turn complex multiplication problems into addition of exponents? People were obviously too lazy to do multiplications of large numbers and wanted to turn them into addition problems of exponents. But two people were lazier than others – John Napier (8th Laird of Merchiston) and Jost Bürgi (a famous Swedish watchmaker) who invented something called logarithms where the entire focus was on exponents. Unfortunately, their original method is too outdated to be used here – and what we will use is the modern rendition of it.

Suppose you have the exponential form $2^3 = 8$, we are asking the question, "What is 2 multiplied to itself 3 times?" and the entire focus is on the answer, which is 8. In Logarithms, we have to shift the focus from the answer to the exponent here. That is to say, now our question will be – "What power of 2 gives us 8?", which is written as follows:

$log_2\, 8 = 3$, which is literally the same as saying $2^3 = 8$.

The word *log* (short for logarithms) simply changes the way in which we frame the question in exponents – apart from that it is literally the same thing. In more general terms, if we write $b^y = x$, then the logarithmic form of the same thing will be $-log_b\, x = y$ (read as log base *b* of *x*) which I can emphasis is same as asking– "what power of *b* will give the value y?" And that's all. Therefore,

A logarithmic function is of the form $y = f(x) = log_a\, x$, if and only if $a^y = x$, where a is a positive number and not equal to 1.

Query 8.4 Based on your knowledge of exponents, discuss why the base *a* of logarithm in the form $y = f(x) = log_a\, x$ can never be negative or 1?

We saw that the graph of exponential functions were of two types – depending upon the base of exponent being decimal or not. The same kind of situation exists in logarithms (this should seem obvious, isn't it? After all, logs are just a different way of dealing with exponents only.)

Illustration 8.3 Let's try to draw graph of $y = f(x) = log_a\, x$, when $a > 1$ and in fact, let's assume the value of *a* to be 2. So the function is $y = f(x) = log_2\, x$.

Let us think what will happen if we put next values of *x* as input. For example, $x = -4$, and $y = f(x) = log_2(-4)$, which is basically asking that what power of 2 will give us -4? That is impossible, no power of 2 would ever give us a negative number, as 2 itself is a positive number. Therefore, we must input only positive real numbers. Drawing a table of input and output values, as shown in Table 8.3.

Table 8.3: Values of the function $f(x) = log_2\, x$

x	1/4	1/2	1	2	4	8	16
$y = log_2 x$	-2	-1	0	1	2	3	4

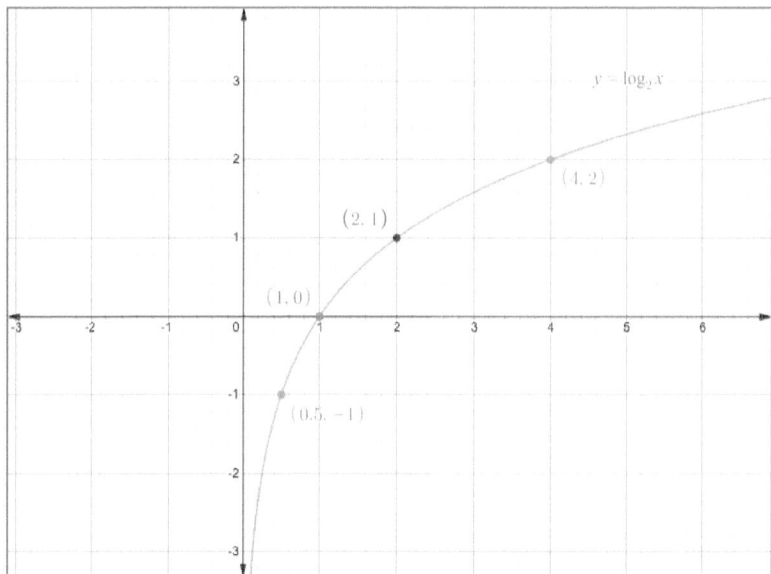

Fig. 8.5: Graph of the function $f(x) = \log_2 x$

Clearly, the graph a vertical asymptote at the y-axis. The domain seems to only positive real numbers (i.e. $x > 0$) and range being all values of R.

Illustration 8.4 Let's try to draw graph of $y = f(x) = \log_a x$, when $0 < a < 1$ (basically its a decimal) and in fact, let's assume the value of a to be 1/2. So, the function is $y = f(x) = \log_{\frac{1}{2}} x$.

Let us think what will happen if we put some values of x as input. For example, $x = -4$, and $y = f(x) = \log_{\frac{1}{2}}(-4)$, which is basically asking that what power of 1/2 will give us -4? That is impossible, no power of 1/2 would ever give us a negative number, as 1/2 itself is a positive number. Therefore, we must input only positive real numbers. Drawing a table of input and output values, we get:

Table 8.4: Values of the function $f(x) = \log_{(1/2)} x$

x	1/4	1/2	1	2	4	8	16
$y = \log_{\frac{1}{2}} x$	2	1	0	−1	−2	−3	−4

Again, the values look like role-reversal cases when the base is not a decimal. Hence, the graph is:

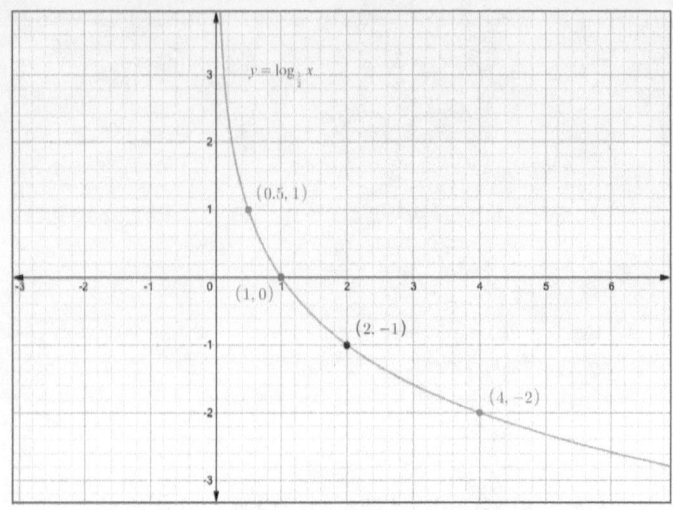

Fig. 8.6: Graph of the function $f(x) = log_{1/2} x$

All conditions are same, vertical asymptote being the y-axis. The Range being R and the domain being $x > 0$.

Query 8.5 What can you say about the graphs of $y = log_{10} x$ and $y = 10^x$?

8.6 A Few Tricks in Logs

There are a bunch of formulas in logarithms that you need to get used to. Of course, they may seem alien at first, but they are just formulas for exponents in a fancier manner. Let's see, if I take two equations, like, $x^a = p$ and $x^b = q$, then the fancier way of writing them would be $a = log_x p$ and $b = log_x q$.

a. If you multiply the two exponents, we get: $x^a . x^b = p.q$ or $x^{a+b} = p.q$ again if we want to write it in fancy logarithmic form, we would get: $log_x (p.q) = a + b$. But $a = log_x p$ and $b = log_x q$, therefore, we can say:

$$log_x (p.q) = log_x p + log_x q$$

(Yet another different of writing the rule of adding the exponents when they are on the same bases multiplied to one another)

b. Here is one for you to prove (because I am too lazy to do all the work):

$$\log_x\left(\frac{p}{q}\right) = \log_x p - \log_x q$$

(it literally goes the exact same way as the previous property, except for one minor change)

c. $\log_1 p = 0$ and $\log_p 1 = 0$ (come on! These two should be super obvious to you)

d. The next one being

$$\log_x p^n = n.\log_x p$$

A super useful formula, the power literally climbs down as a coefficient. Proving this in fact way too simple. As we declared, $x^a = p$ (and of course $a = \log_x p$). If we raise both sides to some power n, we get: $(x^a)^n = p^n$ or $x^{an} = p^n$. Writing this again in logarithmic form, we get: $\log_x p^n = a.n$, but we know already that $a = \log_x p$. Therefore, $\log_x p^n = n.\log_x p$ and that's all!

e. The last one looks complicated but it is quite useful at times. Suppose that you are not happy with the base of logarithm that you are working on, i.e. in $\log_x p = a$, you want some other base y. Then, the way to do it is:

$$\log_x p = \frac{\log_y p}{\log_y x}$$

Would you try proving it? Its simpler than you think.

8.7 The Two Bases of Logs

While we can work with logarithms of any base, two are of special importance to us. One is log with the base 10, also called common logarithms ($log_{10} x$) and another is log with the base e, also called the natural logarithms ($log_e x$). In fact, when you don't write any base to the log, for example $log\ x$, it is understood that the base is 10 and natural logs have a different notation i.e. $log_e x$ is simply written as *ln x*.

Exemplum 8.3: The decrease in population of a town can be modelled by the equation $N = 2000e^{-\lambda t}$, where N is the number of people, t is time in years and λ is a constant. What is the initial population? If the population after 2 years is 1000, then what is the value of λ? Based on this model what is the population of the town after 5 years?

Solution: The initial population can be simply found by putting $t = 0$. We get,

$$N = 2000e^{-\lambda(0)} = 2000e^0 = 2000.$$

Therefore, the initial population of the town is 2000 people. It is given that after 2 years the population is 1000, therefore,

$$1000 = 2000e^{-2\lambda}$$

$$\Rightarrow e^{-2\lambda} = 0.5$$

Taking the natural log of both sides, we get:

$$\Rightarrow \ln e^{-2\lambda} = \ln 0.5$$

$$\Rightarrow -2\lambda.\ln e = \ln 0.5$$

$$\Rightarrow -2\lambda = -0.693$$

$$\Rightarrow \lambda = 0.346$$

Therefore, the equation becomes: $N = 2000e^{-0.346t}$. And the population at the end of 5 years can simply be found out by putting $t = 5$.

Exemplum 8.4 Solve the equation $2\log_2 x - \log_8 x = 10$

Solution: We notice that the bases are not the same here. We can use the base changing formula to make the bases same, which will give us:

$$2\log_2 x - \frac{\log_2 x}{\log_2 8} = 10$$

$2 \log_2 x$ is same as $\log 2x^2$ and $\log_2 8 = 3$, applying these changes, we get:

$$\log_2 x^2 - \frac{\log_2 x}{3} = 10$$

$$\Rightarrow \frac{3\log_2 x^2 - \log_2 x}{3} = 10$$

$$\Rightarrow log_2(x^2)^3 - log_2 x = 30$$

$$\Rightarrow log_2 \frac{x^6}{x} = 30 \text{ [Properties of Log]}$$

$$\Rightarrow log_2 x^5 = 30$$

$$\Rightarrow x^5 = 2^{30} \text{ [Definition of Log]}$$

$$\Rightarrow x = 2^{30/5}$$

$$\Rightarrow x = 2^6 = 64$$

8.8 Closure

Logarithmic and exponential dependencies are not only found in nature – but in a lot of manmade customs too! For example, the pH scale to measure the acidity of a substance is based on common logarithmic scale. The measurement of sound is done in decibel scale is also a logarithmic one! Earthquake is measured with Richter Scale and that too is a logarithmic scale. Exponential functions occur mostly in natural phenomenon like population and bacterial growth, radioactive decay, compound interest, cooling of objects, and growth of phenomena such as virus infections, Internet usage, etc.

Exponential and logarithmic functions will compose a major part in our study of calculus. Unfortunately, they don't behave as nicely as polynomial functions but then who does? Trigonometric functions certainly don't.

You get the picture here. While trigonometric functions describe periodic stuff, these functions help you picturize the growth and decay in nature.

Exercises

1. The Richter scale measure of the magnitude of the earthquake uses the formula

$$R = \log \frac{A}{A_0}$$

   Where, A_0 is the amplitude of the smallest detectable wave (or standard wave) and A is the amplitude of the earthquake wave. The highest rating ever recorded for an earthquake is 9.5 during the 1960 Valdivia earthquake in Chile. Using the definition of the Richter scale, how many times was the earthquake wave stronger than the standard wave?

2. The world's population currently (as of 2019) is 7.7 billion people. It is expected to grow at a rate of 1.08% every year. Form an exponential equation predicting the world population from 2019 growing at this rate. What is the expected world's population in 2025?

3. Prove that $a^{\log_a x} = x, x > 0, a \neq 1$.

4. Solve $ln(x-a) = lnx - a$ (you may leave your answer in terms of e).

5. If $\alpha^x = \beta$ and $\beta^y = \alpha$, then find the value of xy.

6. Find the domain of

   a. $f(x) = \dfrac{4}{\log_{10}(4-x)}$

   b. $f(x) = \log_{10} \sin(x^2)$

   c. $f(x) = \sin^{-1} \log_{10}(x - x^3)$

7. Suppose you hurt yourself while playing and go to a doctor who also happens to be a mathematician. He tells that the area covered by wound will decrease as you heal with time. The area covered by wound can be mathematically modeled by the equation $A(t) = A_0 e^{-0.04t}$, where is the initial area of the wound (say it is 5 square centimeters) and t is the time taken in days. Graph the function and state its domain and range. Will the wound ever heal going by this equation?

8. Solve the equation: $5^{x+4} = 10^{7x-3}$.

9. Graph the function, $f(x) = \dfrac{a}{1+be^{-cx}}$ by taking any values of a, b and c. What can you say about this function when x becomes very large?

10. Newton's law of cooling states that $T = (T_0 - T_s)e^{-\lambda t} + T_s$, where T_0 is the initial temperature of the substance, T_s is the surrounding temperature and λ is the rate of cooling of a substance. A soup is at $80°C$ when it is served to you. It cools at the rate of $0.01°C/s$. How long will it take to come to room temperature?

9 From Language to Grammar: Miscl. Topics in Functions

"Mathematics is a game played according to certain rules with meaningless marks on paper."

— *David Hilbert*

Of all the chapters that we have done so far, this one has the most amount of rigor and technicality. The key here is simple. We have learned some basic functions – or so to say some parts of the language of nature. We now have to move to the Grammar. How to form the language correctly. How to better understand it. A fair warning before you begin – I would like you go really slow with this chapter. Stop at each step and make your own examples to test what you have just studied.

9.1 Moving in Pieces: Piecewise Functions

So far, we have considered functions (polynomials, rational, irrational, trigonometric, logarithmic, exponential) separately and did not try to mix them. This time we are going to spice things up. Let us look at the illustration 9.1.

Illustration 9.1: Let say we define a function as follows:

$$f(x) = y = x, x \leq 0,$$
$$\sqrt{x}, x > 0$$

We have combined a linear polynomial function and an irrational function together. But the catch is, linear polynomial function is valid only for negative values of input ($x \leq 0$) and the irrational function is valid only for positive values of input ($x \geq 0$). Essentially, we have marked the territory ('Stay out of my territory', one function says to another as Walt says to his competitor in Breaking Bad). What can you say about the domain and the range of this function?

The graph of this function is shown in Fig. 9.1. We should already be starting to see the possibilities of defining functions piecewise. For example, the gravitational force exerted by Earth has linear dependence for objects that are below the surface of Earth and it has an inverse square (rational) dependence if the objects is above the surface of Earth. In these kinds of cases, we need to define the functions separately depending on where the input has been provided to it.

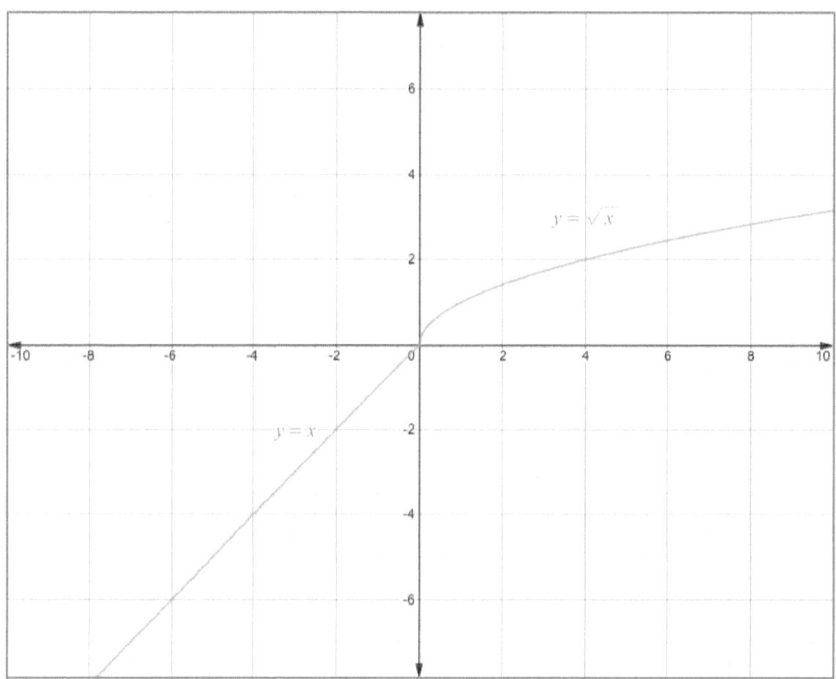

Fig. 9.1: Graph of illustration 9.1

Query 9.1 The acceleration due to gravity of earth for distances below the surface of the earth is given by the formula $g(R) = \frac{4}{3}\pi \rho R$, where ρ is density of the earth (5.51 g/cm^3) and the same for above the surface of earth is given by the formula $g(R) = \frac{GM}{R^2}$, where G is the universal gravitational constant (6.67 × 10^{-11}) and M is the mass of the earth (6 × 10^{24} kg). Write the piecewise function $g(R)$ if the Radius of the Earth is 6400 km and graph it.

9.1.1 Absolute Value Function

One special piecewise function that we will use again and again is the absolute value function (written as $|x|$). This function takes all inputs and simply turns them positive if they are negative. We define it as following:

$$f(x) = |x| = -x, x \leq 0,$$

$$x, x > 0$$

You can see the graph plotted in Fig. 9.2. One can clearly say that the domain of the absolute value function is R but the range is only positive real numbers i.e. $[0, \infty)$.

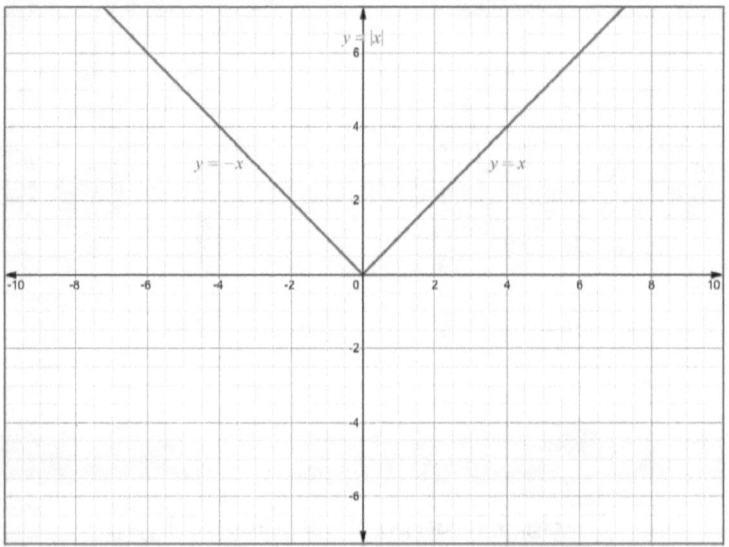

Fig. 9.2: Graph of y = |x|

Exemplum 9.1: Solve the equation $|x - 3| = 4$ and explain the physical meaning of the solution.

Solution: We know that anything you enter in the absolute value function turns positive. In this case, the answer is 4. So the number as an input to the function would have been either +4 or −4 (makes sense, doesn't it?). We consider both the cases.

Case I: $x - 3 = 4 \Rightarrow x = 7$, and

Case II: $x - 3 = -4 \Rightarrow x = -1$

Therefore,

x = −1 *or* 7. Let's focus on the physical meaning of the whole thing. On the number line, both −1 and 7 are 4 units away from 3. So, when we say |x − 3| = 4, we are actually looking for the values of *x* that are 4 units away from 3 in either direction on the number line (kind of looking for distances).

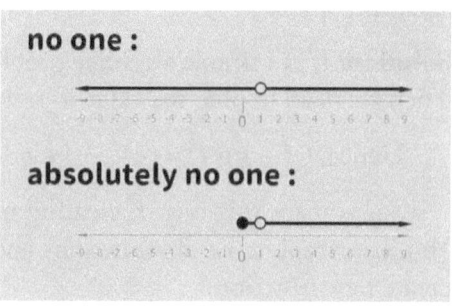

Meme 9.1: Absolutely no one

In general terms, |x − a| = b would actually mean that we are looking for the values of x that are b units from some number a.

Query 9.2 Solve |x − 5| < 7 and explain its physical meaning.

9.2 To Form New Functions

Making pieces of the domain was simply one way of creating more complex functions from the simple one. We have to study three more. Again, remember the basic fact – functions are like primary colors – we mix them to get secondary colors and you mix them to get tertiary colors and somewhere down the line you are able to make a painting. That's what we are doing now. Learning how to paint. Three techniques to get started with i) Algebraic Operations ii) Composition of Functions iii) Inverse Functions.

9.2.1 Algebraic Operations

Many new functions can be created just by adding, subtracting, multiplying and dividing two functions. Let us say there are two functions $f(x)$ and $g(x)$. The following algebraic operations can be done:

1. $f(x) + g(x) = (f + g)(x)$
2. $f(x) - g(x) = (f - g)(x)$
3. $f(x) \cdot g(x) = (f \cdot g)(x)$
4. $\dfrac{f(x)}{g(x)} = \left(\dfrac{f}{g}\right)(x)$

Exemplum 9.2 If $f(x) = x^2$ and $g(x) = \sqrt{x}$, find the values of $(f-g)(4)$, assuming $x > 0$.

Solution: It is a simple algebraic problem. We have $(f-g)(x) = x^2 - \sqrt{x}$. Then we need to plug the value 4 in this expression.

Hence, $(f-g)(4) = (4)^2 - \sqrt{4} = 16 - 2 = 14$.

So, you see this way of creating new functions is not at all complex. You just simply take the functions and do +, −, *, / to these functions to create new functions.

9.2.2 Composition of Functions

So far, we have been putting numbers inside the functions. Let's get a bit more adventurous now. What if we put a function inside another function? For starters, let us take one trigonometric function and one polynomial function. May be, $f(x) = \sin x$ and $g(x) = \sqrt{x}$.

Say we want to put $g(x)$ inside $f(x)$. It will be written as $f(g(x))$ (hope this notation makes sense, it's like saying if we want to put 4 inside a function, we write $f(4)$). Then, $f(g(x)) = \sin g(x)$ or simply say, $f(g(x)) = \sin \sqrt{x}$. What about $g(f(x))$. It will be $g(f(x)) = \sqrt{f(x)}$ or simply say, $g(f(x)) = \sqrt{\sin x}$. Hope you get the difference.

In the standard notation, $fog(x) = f(g(x))$. Look at Fig. 9.3. It shows the mapping diagram of the composition of functions. The function $g(x)$ is mapped from set A to B which means the domain of the function $g(x)$ is the set A and its range is the set B. Similarly, the function $f(x)$ is mapped from set B to C.

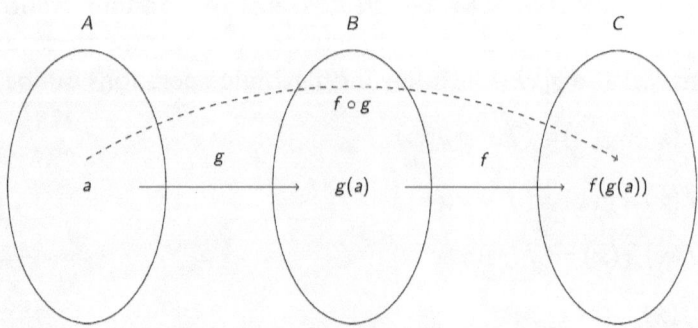

Fig. 9.3: Composition of functions

When we write *f(g(x))*, we are actually creating a direct mapping from set A to set C. An input *x* is fed into *g(x)* and then *g(x)* is used as input for the function *f(x)*. Imagine the possibilities now – all the basic functions that we have studied – polynomial functions, trigonometric etc. all can be fed into each other and we can get more complex functions. That is why it is so important to understand everything about these basic functions. The more complicated functions are generally the composition of these simple functions.

Generally, *f(g(x))* ≠ *g(f(x))*.

Illustration 9.2: One important skill in the future that we will use is the decomposition of the function. To be able to know which simply functions made up the complicated functions that we have. Let's say we have the function:

$$\phi(x) = \log_{10}\sqrt{\sin(x^3)}$$

This is a fairly complicated function. But let's break it one by one-layer by layer. Remember to focus on the basic functions. The first thing we see is a log of something. We call it $f(x) = \log_{10} x$ (or $\log_{10}$ *something*). Next thing is square root of something – an irrational function. We call it $g(x) = \sqrt{x}$. Then, we have a trigonometric function (sin of something), calling it $h(x) = \sin x$. Finally, we have a polynomial function and that we name $i(x) = x^3$. Now, observe the following composition operations.

$$i(x) = x^3$$
$$h(i(x)) = \sin i(x) = \sin x^3$$
$$g(h(i(x))) = \sqrt{h(i(x))} = \sqrt{\sin x^3}$$
$$f(g(h(i(x)))) = \log_{10} g(h(i(x))) = \log_{10}\sqrt{\sin x^3} = \phi(x)$$

So you see the game here!

Query 9.3 Is it true that *h o (f o g) = (h o f) o g* for any three functions *h(x)*, *f(x)* and *g(x)*?

9.2.3 Inverse Functions

The name of this section is a big giveaway. Suppose you have function $y = f(x): X \rightarrow Y$ (means mapped from set X to set Y). So, you put *x* as an input and got *y* as an output when it passes through a function *f*.

When we say inverse of a function, we are looking for a function that maps backwards from set Y to set X. That is to say, it takes y as input and gives the value x as the output. The inverse of a function is denoted by $f^{-1}(x)$.

Clearly if $f(x)$ gives us coordinates (x, y) on the cartesian plane, then $f^{-1}(x)$ will give (y, x) as the coordinates on the cartesian plane. An example is shown in Table 9.1.

Table 9.1: Values of the inverse function

$f(x)$	(2,4)	(3,6)	(4,8)	(5,10)	(6,12)	(7,14)	(8,16)
$f^{-1}(x)$	(4,2)	(6,3)	(8,4)	(10,5)	(12,6)	(14,7)	(16,8)

That brings us to the question on how to actually find the inverse of a function. Let us look at Illustration 9.3 to understand this.

Illustration 9.3 Suppose we are working with this function $y = f(x) = 4x + 3$. We wish to find the inverse function of this. Let us first analyze what happens to x after entering this function. It first gets multiplied by 4 followed by an addition with 3. So what would be the inverse procedure of this in order to get back the variable x. Simple!

Typically, we should subtract the 3 we added and then divide by 4 to get back x. In symbols, $f^{-1}(x) = y = \dfrac{x-3}{4}$ should act as an inverse function. Is it true though? Let us check. We will see what we get by putting $x = 1$ in the function $f(x)$.

$f(1) = 4(1) + 3 = 7$

Now, the inverse function should take 7 as input and give back the value we started out with i.e. 1. Let's check.

$f^{(-1)}(7) = \dfrac{(7-3)}{4} = 1$ and bingo! It works. I must admit that guess the inverse function for this illustration was way too simple. However, in most cases, it won't be this simple. We shall apply the following technique:

a. Replace x with y and y with x in the function $f(x)$

b. Then solve for y to get $f^{-1}(x)$.

Let us see if this works with the function in this illustration i.e. $f(x) = y = 4x + 3$. In order to find the inverse:

a. $x = 4y + 3$ (Replacing x with y and y with x in the function $f(x)$)

b. $4y = x - 3 \Rightarrow y = \dfrac{x-3}{4}$ (Solving for y)

And there it is! We get the exact same answer. This technique works with most kinds of functions.

But hold on! We are not done yet! Inverse of a function is a very tricky thing. Just because we can now find the inverse, doesn't mean inverse will exist at all times. Say for example, the function . Let us try to find the inverse of this function $f(x) = x^2$ in the following Illustration.

Illustration 9.4 We try to find the inverse of the function $f(x) = x^2$.

We can make a list of values of the function $f(x) = x^2$

Table 9.2: Values for $f(x) = x^2$

x	−3	−2	−1	0	1	2	3
$y = x^2$	9	4	1	0	1	4	9

Now, we shall try to find the inverse function for $f(x) = x^2$. We basically square the value of x here. So the inverse process would be to take the square root. Therefore, $f^{-1}(x) = \sqrt{x}$. We interchange the rows in Table 9.2, we can actually get the table of values for $f^{-1}(x)$.

Table 9.3: Values for $f^{-1}(x) = \sqrt{x}$

x	9	4	1	0	1	4	9	−1
$f^{-1}(x) = \sqrt{x}$	−3	−2	−1	0	1	2	3	undef.

Before telling you the problem here, let me remind you of the image (Fig. 9.4) that you saw in Chapter 4 about what actually counts as a function.

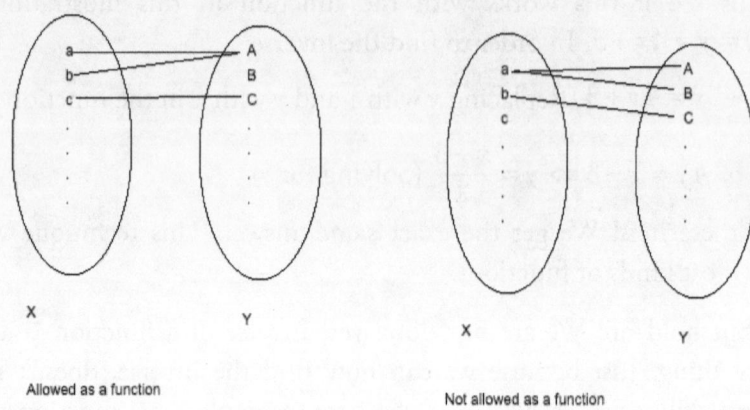

Fig. 9.4: What is and what is not a function

Notice that 9 in the domain is mapped to both 3 and −3, and so is 4 mapped to 2 and −2. Also there is no root for negative values. Remember what we studied in the chapter functions?

a. *Every element in X must be mapped to a unique element in Y. (one-one mapping)*

b. *Every element in X must have a mapping. (onto mapping)*

Both of these conditions are not satisfied here! Neither the elements in input are mapped to unique values nor every value in domain has a mapping! So technically $f^{-1}(x)$ is not even a function! What to do now?

The problem in Illustration 9.4 can be solved simply by **restricting the domain**. Which simply means that we will remove the values from the domain that create problems for us. In the inverse obtained in Illustration 9.4, if we remove the negative numbers and make the domain as $x \geq 0$, then the problem is solved!

When our function $f(x)$ is not one-one, then we will have to restrict the domain while finding the inverse in order for the inverse function $f^{-1}(x)$ to actually make sense.

Exemplum 9.3: Find the inverse of the function $f(x) = x^3$ and graph it.

Solution: Clearly, the inverse of the function $f(x) = x^3$ is simply $f^{-1}(x) = x^{1/3}$. The graph is shown in Fig. 9.5.

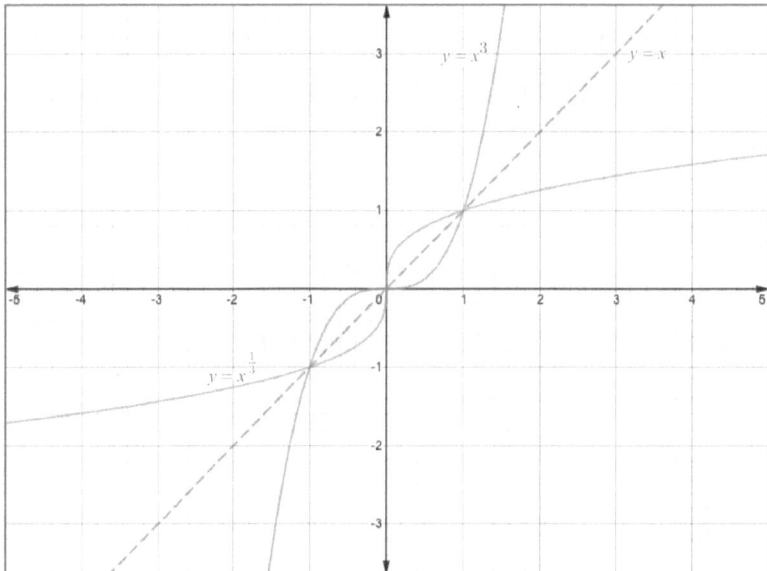

Fig. 9.5: x^3 and its inverse

Notice one thing in Exemplum 9.3, **the graph of the inverse function looks like it is a reflection of the original function about the line $y = x$.** That is a common property of all functions and their inverses. Can you tell why?

Query 9.4 If $f(x)$ is a function, then do $f^{-1}(x)$ and $f(x)^{-1}$ have same meaning?

Query 9.5 Is it true that $f^{-1}(f(x)) = f(f^{-1}(x)) = x$? What is the meaning of it?

9.2.3.1 Inverse Trigonometric Functions

In the last section, we discussed how if the functions are not one-one and onto, there is a problem with the inverses. Well, what better example of this than the naughty trigonometric functions! For example, sine and cosine function repeats its value after every 2π radians, tangent function repeats its value after every π radians. Definitely the inverse of trigonometric functions are not functions. Remember the Bernoullis from Chapter 3? Daniel Bernoulli was one of the first guys to consider inverse trigonometric functions.

For example, we know that $\sin\frac{\pi}{6} = 0.5$, but if we ask sine of what angle is 0.5 i.e. $\sin^{-1} 0.5 = ?$ (John Herschel introduced the $\sin^{-1}$ and $\tan^{-1}$ notations in an article in the Philosophical Transactions of London in 1813). Well, there are infinite answers to this.

Therefore, we restrict the range of these inverse trigonometric functions so that one input must not give multiple outputs. The restriction of ranges (or domain of the original trigonometric functions) of inverse trigonometric functions are shown in Table 9.4.

Table 9.4: Domain and Range for inverse trigonometric functions

Function	Domain	Range
$\sin^{-1}(x)$	$[-1,1]$	$\left[-\frac{\pi}{2}, \frac{\pi}{2}\right]$
$\cos^{-1}(x)$	$[-1,1]$	$[0, \pi]$
$\tan^{-1}(x)$	$(-\infty, \infty)$	$\left(-\frac{\pi}{2}, \frac{\pi}{2}\right)$
$\csc^{-1}(x)$	$R - [-1,1]$	$\left[-\frac{\pi}{2}, \frac{\pi}{2}\right] - \{0\}$
$\sec^{-1}(x)$	$R - [-1,1]$	$[0, \pi] - \{\frac{\pi}{2}\}$
$\cot^{-1}(x)$	$(-\infty, \infty)$	$(0, \pi)$

Note that these ranges are not randomly selected. For example, $\sin^{-1} x$ has a range of $\left[-\frac{\pi}{2}, \frac{\pi}{2}\right]$, which basically means 1st and 4th quadrant. Now, sine function is any positive in 1st and 2nd quadrant, but it is positive in 1st and negative in 4th quadrant – thus taking care of all sorts of values. The same stands for all other ranges.

Exemplum 9.4: Prove that $\sin^{-1} x + \cos^{-1} x = \frac{\pi}{2}$

Solution: By writing $sin^{-1}x$, we mean to ask if there is some angle whose sine value is x. Let us call that angle as θ.

Then,

$$sin^{-1}x = \theta \ldots \text{(i)}$$
$$\Rightarrow x = sin\,\theta$$
$$\Rightarrow x = cos\left(\frac{\pi}{2} - \theta\right)$$
$$\Rightarrow cos^{-1}x = \frac{\pi}{2} - \theta$$
$$\Rightarrow cos^{-1}x = \frac{\pi}{2} - sin^{-1}x \text{ [From (i)]}$$
$$\Rightarrow sin^{-1}x + cos^{-1}x = \frac{\pi}{2}$$

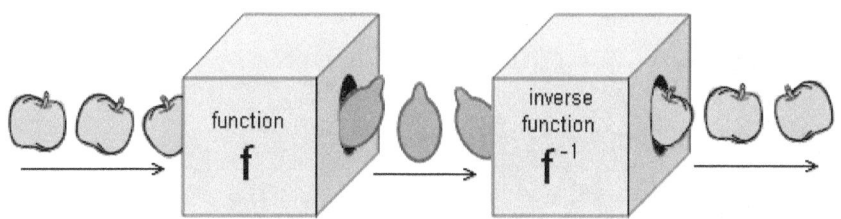

Fig. 9.6: Inverse Functions

9.3 Closure

First of all, the sheer length of this topic is insane, but I decided to give you the main ideas rather than elaborate and rigorous concepts. The goal is not to show how difficult this chapter can be, but rather telling you the basic idea in the simplest manner possible. Remember that while we do need to master functions, our main goal is towards Calculus. Having said that, I strongly advise you to go through each topic in detail that are given over here – especially inverse trigonometric functions. Composition of functions and inverse functions is something that we will encounter again and again in Calculus.

Exercises

1. A company makes basketballs of 600 grams. An error of 30 grams is tolerable in the weight. Write an absolute value inequality describing the weights of the basketballs that are acceptable.

2. The sigma function is defined by $y = \dfrac{|x|}{x}$. Find the domain and range of this function and graph it.

3. Find the inverse of the following functions:

   a. $f(x) = \sqrt[5]{\dfrac{2-x}{x}}$

   b. $f(x) = x^2 + 5x + 3$

   c. $f(x) = \ln x^2$

   d. $f(x) = \dfrac{ax+b}{cx+d}$

   e. $y^3 + \cos^2 x - y + a = 0$

4. Find domain of the following functions:

   a. $f(x) = \sqrt{\log_4\left(\dfrac{2-x^3}{5}\right)}$

   b. $f(x) = \cos^{-1} \dfrac{2}{2 + \cos 2x}$

   c. $f(x) = \dfrac{|x|}{\sqrt{x - 2|x|}}$

   d. $f(x) = \log_3 \cos x^3$

   e. $f(x) = \sin^{-1} |1 - x^2|$

5. Let $g(x) = \sqrt[7]{y - x^7}, x > 0$. Find the value of $g(g(x))$.

6. Find the range of the function $\phi(x) = |\sin x|$.

7. Show that the function $f(x) = \beta + \dfrac{1}{x - \beta}$ is its own inverse.

8. If a horizontal line crosses the graph of a function two times, what can you say about its inverse?

9. Consider two functions $f(x) = x^2 + 5x + 6$ and $g(x) = |x|$. Solve the inequality $g(f(x)) \leq 12$.

10. If the domain of $f(x)$ is (a, b) and the domain of $g(x)$ is (b, c), what is the domain of $(f + g)(x)$?

10 The Mathematical form of Art: Graphing Functions

"I am interested in mathematics only as a creative art"
– G.H. Hardy

10.1 Closure

Surprisingly, unlike all other chapters, closure comes here first. It is because in a way it is. We have learned several basic kinds of functions. We will now explore how they can be transformed into different other complicated forms. We call these are *transformations of a function*. This chapter here is more of a work of art – as the entire mathematics is. You may see that we are back to our tree of functions now, the only difference is that now, it's more decorated.

Fig. 10.1: The updated tree of functions

Being able to graph the functions is a key to understanding how it behaves – something very crucial for our study in Calculus. What we need now is simply a table of transformations – something that will tell us what happens to the graph of a function when we perform different operations on it. That's all!

Table 10.1: A guide to transformations of functions

Transformation	What it does?
$f(x + h)$	This shifts the graph of the original function $f(x)$ to the left by h units ($h>0$).
$f(x - k)$	This shifts the graph of the original function $f(x)$ to the right by k units ($k>0$).
$f(x) + a$	This shifts the graph of the original function $f(x)$ upwards by a units ($a>0$).
$f(x) - b$	This shifts the graph of the original function $f(x)$ downwards by b units ($b>0$).
$af(x)$	This stretches the graph of the original a times along the y-axis.
$f(ax)$	This stretches the graph of the original a times along the x-axis.
$|f(x)|$	This reflects the part of the original graph $f(x)$ lying below the x-axis along the x-axis.
$f(|x|)$	This ignores the part of graph for negative values of x and reflects the rest along y-axis.
$f(-x)$	This reflects the graph of the original function $f(x)$ along the y-axis.
$-f(x)$	This reflects the graph of the original function $f(x)$ along the x-axis.

Illustration 10.1: Let us try to graph the function $f(x) = (x + 3)^2 + 2$. We shall deal with each transformation, one by one, starting from the basic function. That's the key.

First of all, let's identify the basic function. It seems to be $f(x) = x^2$. We already know the graph of $f(x) = x^2$ as shown in Fig. 10.2. It has a vertex at (0, 0). The next step is $f(x) = (x + 3)^2$ which shifts the graph towards left by 3 units. Thus the new vertex is at (0,–3). The transformation is shown in Fig. 10.3. The final step is $f(x) = (x + 3)^2 + 2$ which shifts the graph by 2 units upwards and finally the vertex is at (–3, 2). That completes the graph we were looking for as shown in Fig. 10.4

188 ■ The Calculus Experience

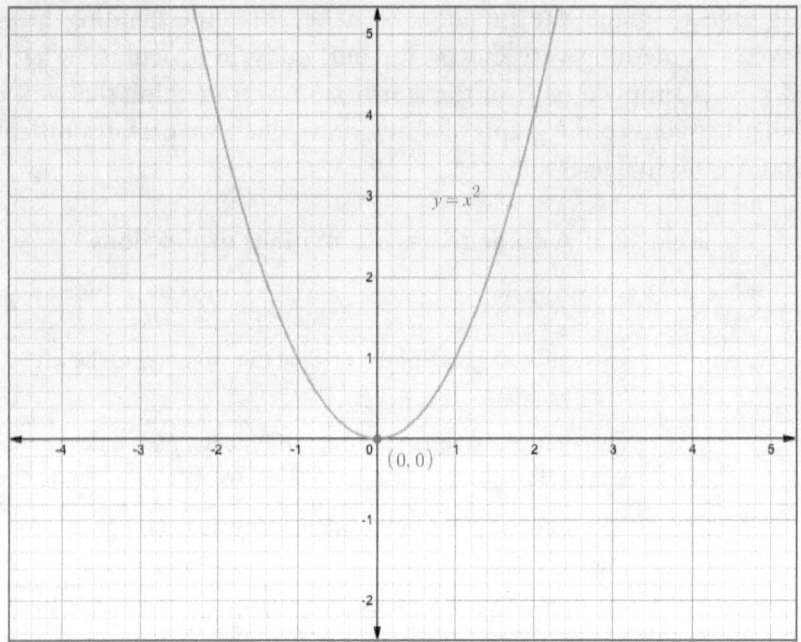

Fig. 10.2: Graph of $f(x) = x^2$

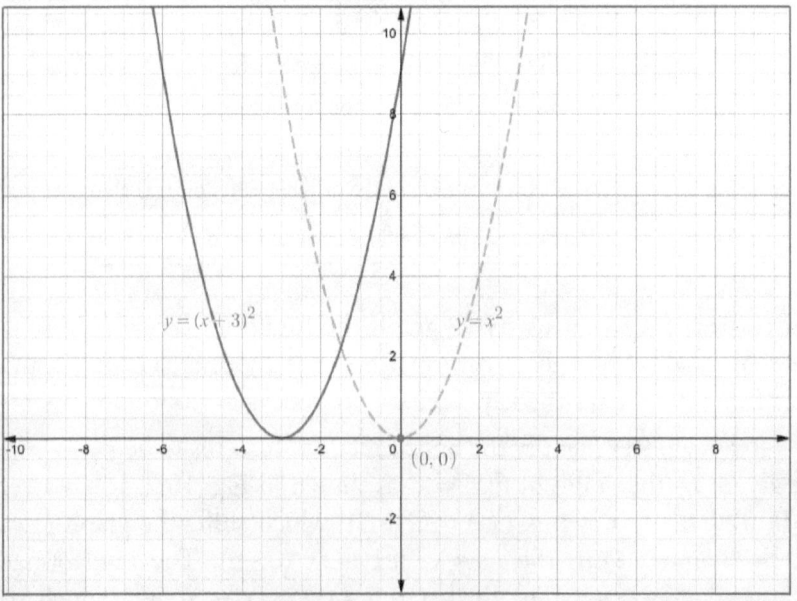

Fig. 10.3: Graph of $f(x) = (x + 3)^2$

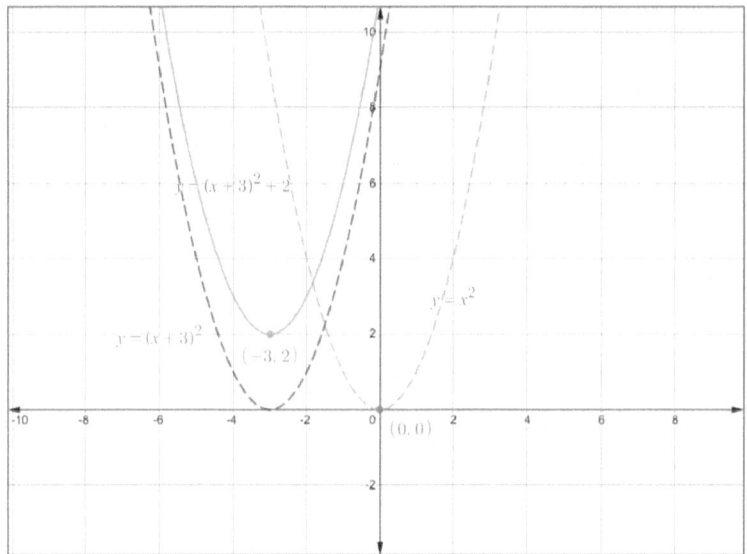

Fig. 10.4: Graph of $f(x) = (x + 3)^2 + 2$

Illustration 10.2: Graphing $f(x) = -|\sin x| + 1$. Again, let's start step by step. The basic function here is the trigonometric function $f(x) = \sin x$, the graph which is shown in Fig. 10.5.

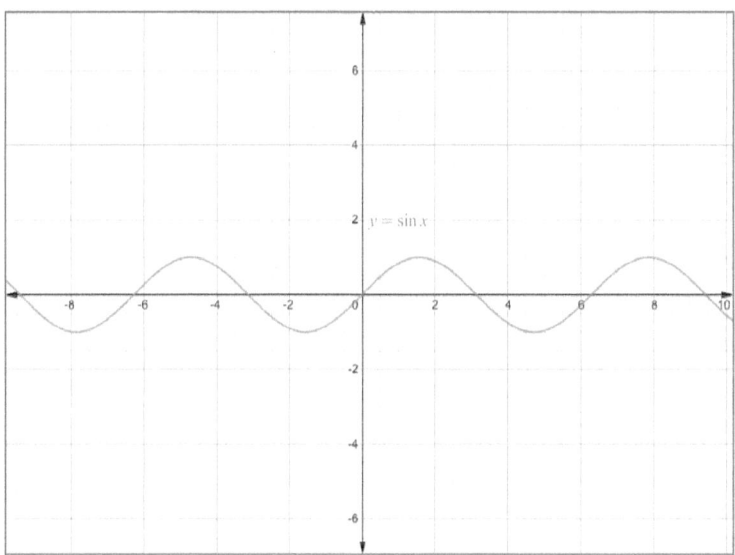

Fig. 10.5: Graph of $f(x) = \sin x$

The next step is $f(x) = |\sin x|$. And what does it do? It reflects only the part of the graph lying below x-axis along the x-axis as shown in Fig. 10.6. Then, $f(x) = -|\sin x|$, according to the table will reflect the entire graph about x-axis as shown in Fig. 10.7. Now the final step is simple $-f(x) = |\sin x| + 3$, it simply shifts the previously obtained graph in Fig. 10.7 upwards by 3 units as shown in Fig. 10.8. That's all!

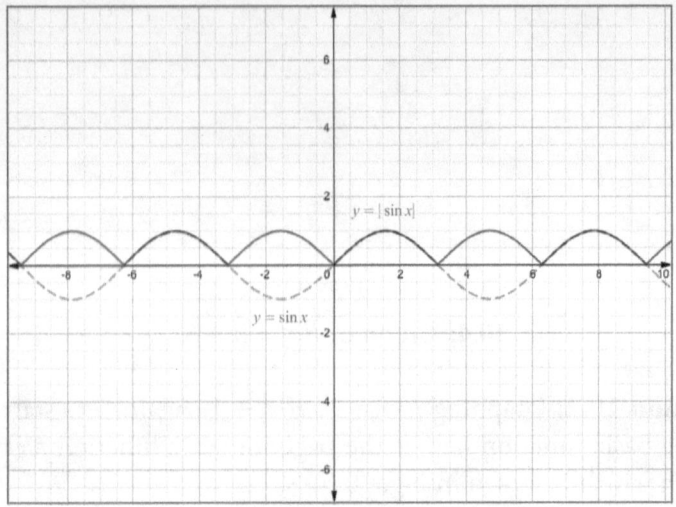

Fig. 10.6: Graph of $f(x) = |\sin x|$

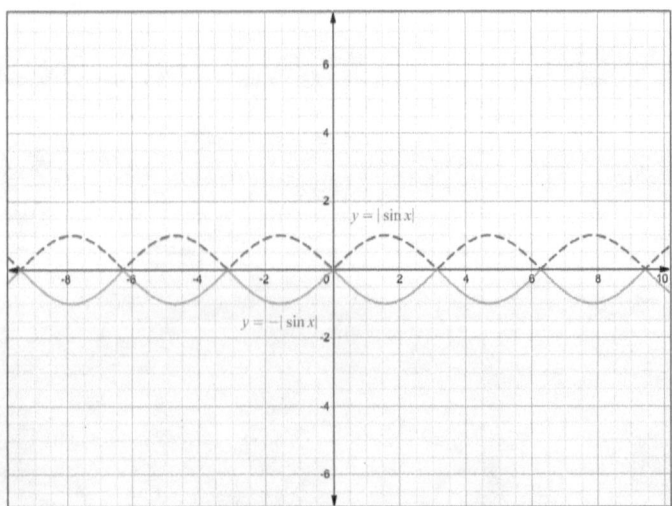

Fig. 10.7: Graph of $f(x) = -|\sin x|$

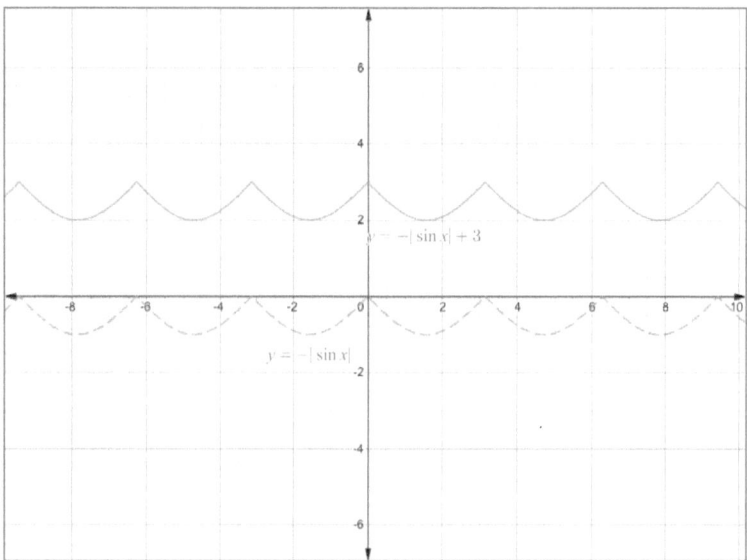

Fig. 10.8: Graph of $f(x) = -|\sin x| + 3$

Illustration 10.3: Draw the graph of the function $f(x) = e^{|x|} + 4$. By now it must be obvious that the starting function is the basic function, which in this case is the exponential function $f(x) = e^x$. The graph of which is shown in Fig. 10.9.

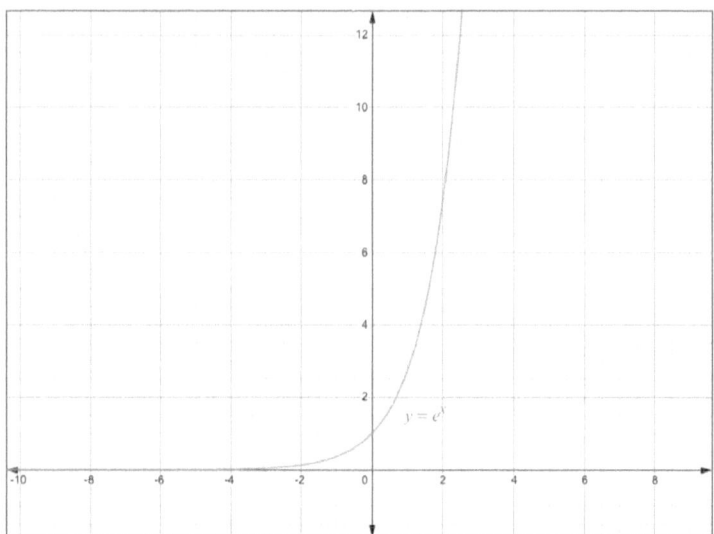

Fig. 10.9: Graph of $f(x) = e^x$

The next step is $f(x) = e^{|x|}$ which cuts down the graph for the part where x values are negative and reflects the rest part along the *y-axis* as shown in Fig. 10.10. Finally, $f(x) = e^{|x|} + 4$ simply shifts the entire graph upwards by 4 units. The final diagram is in Fig. 10.11.

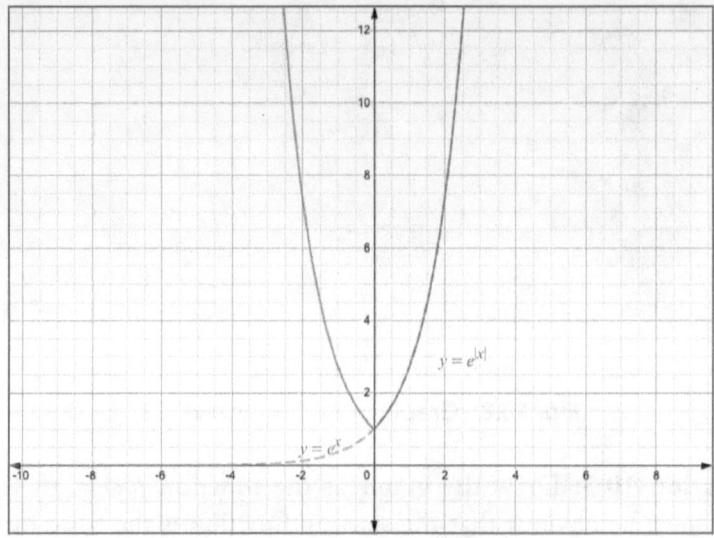

Fig. 10.10: Graph of $f(x) = e^{|x|}$

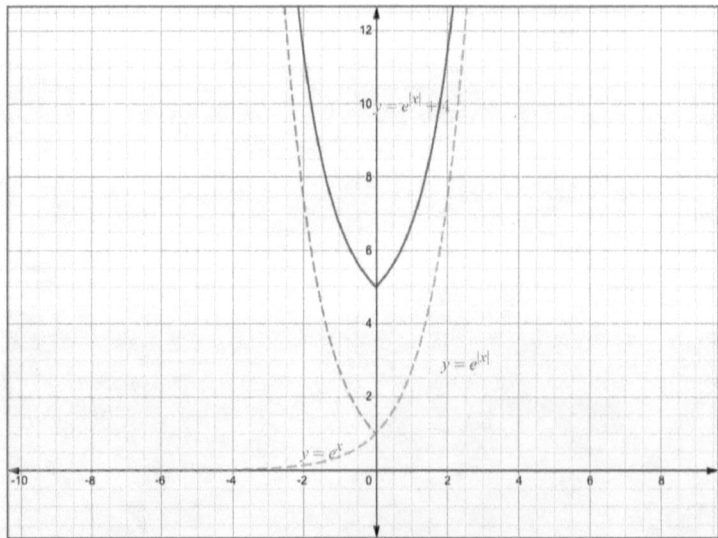

Fig. 10.11: Graph of $f(x) = e^{|x|} + 4$

Illustration 10.4: As a final illustration let's look at the graph of the function $f(x) = 2\sin\left(2x + \dfrac{\pi}{2}\right) + 2$. Looks complicated, but once again – we shall attempt it step by step. The first step is any way the basic function $y = \sin x$ as shown in Fig. 10.12. The next is $f(x) = 2\sin x$, which basically tells us to stretch the entire graph along the y-axis and as a result the new amplitude is now 2 as shown in Fig. 10.13

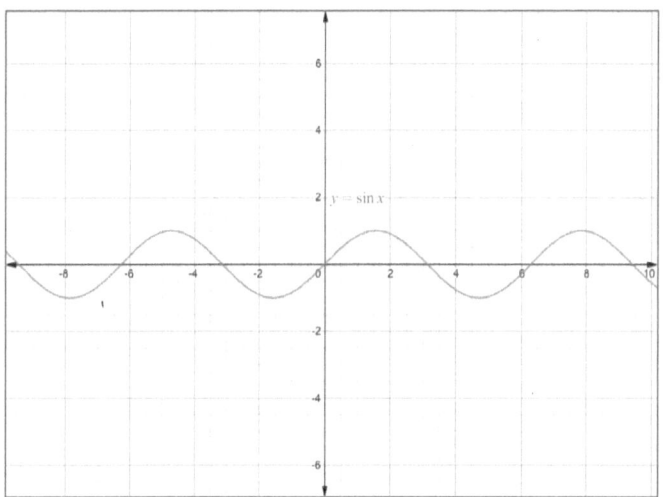

Fig. 10.12: Graph of $f(x) = \sin x$

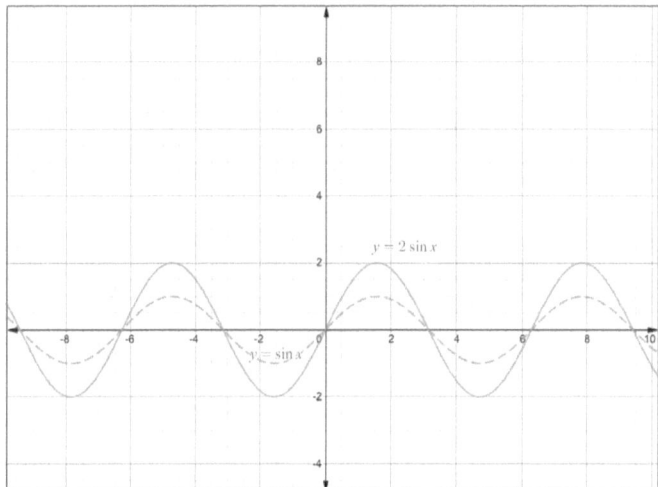

Fig. 10.13: Graph of $f(x) = 2\sin 2x$

Now, $f(x) = 2sin\ 2x$ simply stretches the previous graph again, but this time along x-axis, that means the period of previous function was 2π, but this time it will be just as shown in Fig. 10.14. The second last step is simply $f(x) = 2sin\left(2x + \dfrac{\pi}{2}\right)$ will simply result in a horizontal translation of the whole graph by $\dfrac{\pi}{2}$ units towards the left (doesn't it resemble cosine graph now) as shown in Fig. 10.15

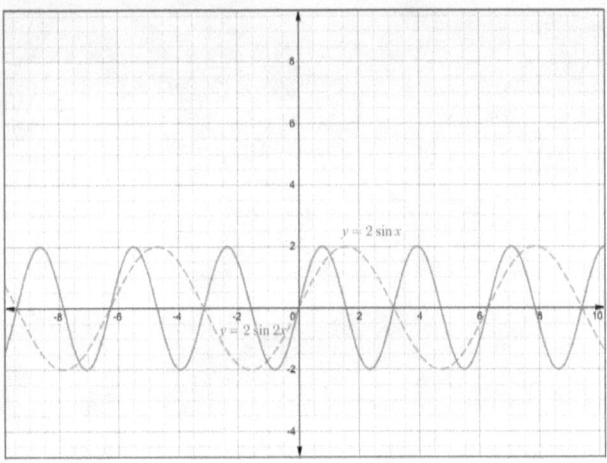

Fig. 10.14: Graph of *f* (x) = 2sin 2x

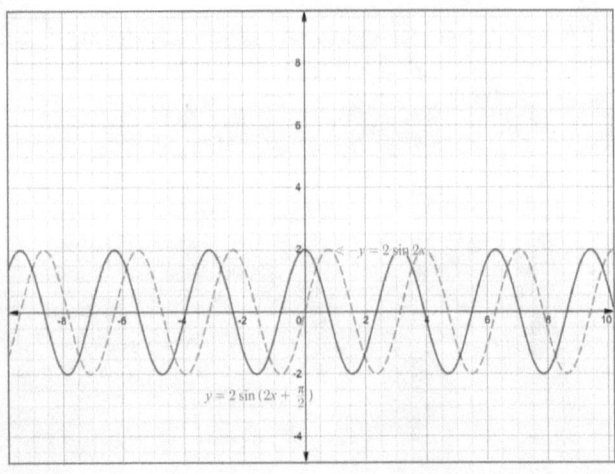

Fig. 10.15: Graph of $f(x) = 2sin\left(2x + \dfrac{\pi}{2}\right)$

The last step should not be obvious to you – $f(x) = 2\sin\left(2x + \dfrac{\pi}{2}\right) + 2$ simply shifts the last graph 2 units upwards. And there we go! By dealing with each transformation one by one, we have successful drawn this graph as shown in Fig. 10.16

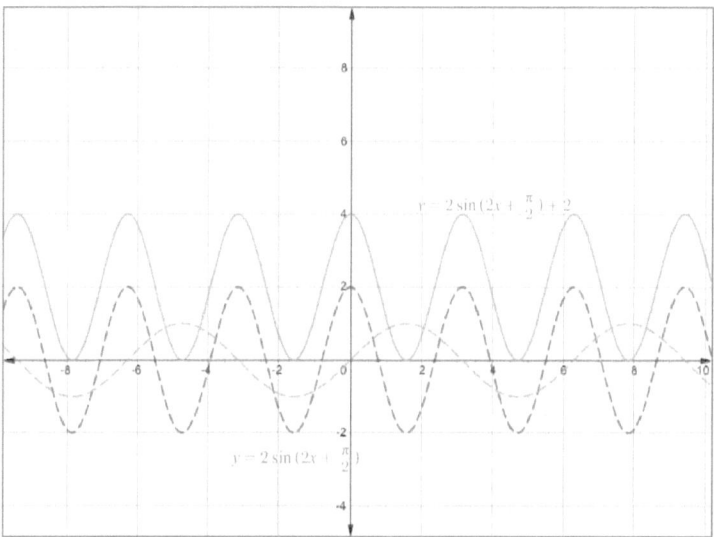

Fig. 10.16: Graph of $f(x) = 2\sin\left(2x + \dfrac{\pi}{2}\right) + 2$

10.2 Half-Knowledge

We can now draw a fair deal of graphs – however, it is just the beginning. The real game will begin in the next book. Hence, what we have obtained here is just half-knowledge. And we all know, half-knowledge is dangerous. I hope you have enjoyed brushing up the basics of Calculus known as precalculus so far.

There is a reason that I have not loaded with you huge number of problems and examples in this entire book. I never wanted to impose what Richard Feynman called the "rigidity of thought". To quote Feynman, "We must leave freedom for the mind to wander about in trying to solve the problems.... The successful user of mathematics is practically an inventor of new ways of obtaining answers in given situations. Even if the ways are well known, it is usually much easier for him to invent his own way— a new way or the old way— than it is to try to find it by looking it up."

The goal is not to train you in mathematics – but rather you get the will to pursue the subject on your own will. Trust me – thousands of books, including this one, will never be able to teach you what you shall learn on your own will. Peace out.

Exercises

Graph the following functions:

1. $f(x) = |\sin|x| + 4|$

2. $f(x) = \dfrac{|\sin x|}{\sin x}$

3. $f(x) = 2\tan(x + \pi) + 4$

4. $f(x) = \left|\dfrac{1}{x^2}\right| - 5$

5. $f(x) = \sqrt{x+3} - 7$

6. $f(x) = \sqrt[5]{|x| + 2}$

7. $f(x) = 3|\ln(x-3)|$

8. $f(x) = e^{x-4} + 1$

9. $f(x) = -\cos(-x) + \pi$

10. $f(x) = \dfrac{4}{5x^{3/2} + 9}$

Also, find the graphs of the inverse of each function above.

11. What is the difference between $(-x)$, $-f(x)$, $f(x)^{-1}$ and $f^{-1}(x)$?

References

1. Tarasov, L.V. (1982). *Calculus: Basic Concepts for High School.* Moscow: Mir Publishers.

2. Stewart, James (2015). *Calculus: Early Transcendentals, 8th Edition.* Cengage Learning.

3. Boyer, Carl. B. (1968). *A history of Mathematics.* New York. John Wiley & sons.

4. Wildberger, N.J. (2016). *Astronomy and Trigonometry in India.* [Video File]. Retrieved from: https://www.youtube.com/watch?v=2HmMzhZ8zJg&list=PL55C7C83781CF4316&index=39.

5. Wildberger, N.J. (2011). *Infinity in Greek Mathematics.* [Video File]. Retrieved from:https://www.youtube.com/watch?v=mKgt1K6VU0k&index=5&list=PL55C7C83781CF4316.

6. Lemmermeyer, Franz (2003, December 21). *Some problems of Diophantus.* Retrieved from: http://www.fen.bilkent.edu.tr/~franz/M300/diopro.pdf.

7. Cajori, Florian (1894). *A history of Mathematics.* New York. McMillan and Co.

8. Boyer, Carl. B. (1949). *The history of Calculus and its conceptual development.* New York. Dover Publications.

9. Hoffman, Johan & Johnson, Claes & Logg, Anders (2004). *Dreams of Calculus: Perspectives on mathematics education.* New York. Springer.

10. Joseph, George. Gherverghese. (1991). *The Crest of the peacock: Non-European roots of mathematics.* London. Princeton University Press.

11. Usiskin, Zalman, Peressini, Anthony, Marchisotto, Elena, and Stanley, Dick. (2003). *Mathematics for High School Teachers: An Advanced Perspective.* Prentice Hall, Upper Saddle River, NJ.

12. Katz, Victor, and Michalowicz, Karen (ed.), Historical Modules for the Teaching and Learning of Mathematics. The Mathematical Association of America. Washington, DC. 2005.

www.ingramcontent.com/pod-product-compliance
Lightning Source LLC
Chambersburg PA
CBHW020906180526
45163CB00007B/2641